THE GEOMORPHOLOGY

OF THE

COTSWOLDS

by

Andrew Goudie and Adrian Parker

School of Geography,
University of Oxford,
Mansfield Road,
Oxford OX1 3TB.

THE
COTTESWOLD NATURALISTS' FIELD CLU
(Founded in 1846)

"THE GEOMORPHOLOGY OF THE COTSWOLDS"

The Cotteswold Naturalists' Field Club is honoured to publish this new and learned work by Professor Goudie and Doctor Parker.

Whilst over the years many interesting and original papers have been printed in our annual Proceedings, which have appeared without a break since our foundation in 1846, this is the first book to bear our imprint since "The Flora of Gloucestershire", published in 1948.

Bearing in mind that this year is our sesquicentennial anniversary, it is particularly appropriate that we are able to support such a fine work of scholarship.

My Council and I compliment the joint authors and wish them well.

Horace Harris,
President.

1st November, 1996

Dr Robert P. Beckinsale M.A., D. Phil
Honorary Member and former Editor of the *Proceedings*

Contents

List of Figures

List of Tables

Chapter 1

1.1 Introduction: a brief history of geomorphology in the Cotswolds

The Cotswolds or Cotteswolds, were the scene of some of the most prolific and important geological research in the early days of that science. William Smith, for example, the 'Father of English Geology', was born in 1769 in the Cotswold village of Churchill, near Stow-on-the-Wold. He served his apprenticeship to a surveyor at Stow itself, and then set up in practice for himself at Bath, in the southern Cotswolds. He died at Northampton in 1839 and so was buried on the Oolite as he had been born on it.

However, the names of Buckman, Murchison, Woodward, Witchell, Lonsdale, Hull and Phillips (Smith's nephew), so illustrious in the the general story of British geology, were also intimately connected with the Cotswolds. These workers, more fluent with the pen than their 20th century successors, were mainly concerned with the palaeontology and stratigraphy of the Mesozoic rocks rather than with the physiography or geomorphology and this trend has been perpetuated to the present. Occasionally a geologist like the Revd. W.D. Conybeare (e.g. 1823) might comment at more than usual length on the deep valleys that fretted the Oolitic mass. In this century Arkell, Ager and Richardson have continued geological research though with little emphasis on landforms. Linsdall Richardson has perhaps done the most work on landforms through his interests in hydrology, and his writings, spread almost over 70 years, must form the basis of nearly all research in the area. He died on New Years Day, 1967. It was through his prominent position on the editorial committee of the Cotteswold Naturalists' Field Club that the *Proceedings* became a major source of geological information of both a local and wider interest.

Whilst it would be idle to maintain that the area has been completely neglected by geomorphologists, the number of papers published from this region, when compared to other regions of Britain, remains poor. However, William Morris Davis, the great American geomorphologist, to the delight and slight bewilderment of the Field Club, made a lightning visit to the area in the 1890s in the company of S.S. Buckman. Together they worked out a theory on the development of rivers, and particularly the genesis of the Severn, but they did not concern themselves with the detection of erosion surfaces. Essentially they examined the evidence for river capture and misfit valleys.

Perhaps the most active geomorphologist to have worked extensively in the mid part of this century has been Robert Beckinsale. He was born on the 23rd July 1908 in Stanford-in-the-Vale, formally in Berkshire, now Oxfordshire. In 1933, then of 'The Leaze', Burford, Oxon, he was elected as a member of the Field Club on Tuesday 20th June, 1933, having been proposed by Prof. H.L. Hawkins and seconded by Mr. Roland Austin (Knight and Riddelsdell, 1933, p.15 *PCNFC*, **XXV** (I)). In the early 1930s Beckinsale had undertaken research into possible or probable boundaries to the Cotswolds with the help and encouragement of Linsdall Richardson (Beckinsale, 1934). As will be seen later, their views on this matter were not always in total agreement, though they always remained great friends. Beckinsale went on to publish numerous papers on the Cotswolds and was one of the first persons to apply geographical and geomorphological thought in order to explain the physiography of this region.

Perhaps rather anomalously the whole area has been little affected by those geomorphologists seeking to reconstruct long term erosional development (the denudation chronologists), particularly when one compares the Cotswolds with the neighbouring regions of Wales, the Forest of Dean, or Southern England. Indeed Beckinsale and Smith (1955) concluded that the area was not suited to the techniques used by denudation chronologists. However, since the 1960s as geomorphology became less concerned with landform evolution and more concerned with process studies, morphometry, and systems, considerable dissatisfaction has been expressed about denudation chronology as a basis for the discipline of geomorphology (Chorley, 1965).

The same rivers studied by Davis and Buckman became the subject of general theories of drainage development in southern Britain, propounded by Lake (1915) and others. Smith and Beckinsale examined the dip-slope rivers for knick points by the use of detailed survey in the early fifties, though much of their work remains unpublished. G.H Dury augered and surveyed the valley bottoms of just those same streams Davis had visited almost sixty years before him. Dury, with some degree of academic feeling against the late Professor of Physical Geography at Harvard, substituted a theory of great hydrological change in the late Pleistocene for theories of river capture in order to account for the misfit valleys (Dury, 1955, 1958, 1965).

A further anomaly in the history of Cotswold geomorphology was the concentration by many workers on mass movements on slopes. The sub-aerialists of the nineteenth century, of whom Witchell was the most competent, used the large landslips encountered in the valleys of the wool-manufacturing district as powerful evidence in support of their theories. In 1867 Witchell wrote,

> "...in the valleys of the Cotteswolds there is abundant evidence of cenudation, but so far as research as hitherto extended, it appears to be subaerial in character, facilitated and increased by landslips."

In a later work Witchell (1882) gives numerous examples of landslips with their dates and the area of the ground involved;

> "...numerous small slips occur every year, one slip which extended some two hundred yards along the hillside, occurred at Cowcombe Hill; and another near Lilleyhorn, Bisley, covered three acres...near Nailsworth in 1826...an area of five acres moved forward in the course of a single night."

The only other major field of research has been the examination of the area's Pleistocene geology and geomorphology. Dury (1951) searched for the shoreline of Lake Harrison in the neighbourhood of Bredon Hill, and Arkell (1947a) and Shotton (1960) have searched for polygonal forms in nearby deposits. Sandford was the first person to assess critically the gravel terraces of the Thames (1924, 1926, 1929, 1954, 1965), while Beckinsale and Richardson (1964) and Tomlinson (1941) undertook detailed stratigraphic work on the deposits at the scarp foot.

Since the 1970s, with the paradigm changes in geomorphological thought, the Cotswolds has witnessed an increase in research in the fields mentioned above. The purpose of this work is to review the geomorphological research from the Cotswolds which includes new research ideas and data which we hope shall prove both interesting and stimulating for others. It is with this in mind that we have dedicated this work to the lifetime achievements in geomorphology of Dr. Robert Percy Beckinsale, whose own work has led to new developments and thought in this field, especially in the Cotswolds, an area which offers so much but has been studied by so few.

1.2 Delimitation of the Cotswolds

The precise limits of the Cotswolds or Cotswold Hills (formerly Cotteswold) has been the subject of much debate since the nineteenth century (Buckman, 1903; Richardson 1935, 1946; Beckinsale, 1992).

S. S. Buckman, a skilled geologist and prominent topographer, based his ideas on the limits of the Cotswolds on traditional opinion, geology, relief and surface appearance of the landscape (Beckinsale, 1992, p. 24). They were published, together with a map, in the *Proceedings of the*

Cotteswold Naturalists' Field Club, 1903, Vol. **XIV**, pt.3, pp.205-242 entiltled "The Cotteswold Hills: A Geographical Enquiry."

Beckinsale (1992,p.22-24) re-published the full account of Buckman's delimitation of the Cotteswold Hills which shall again be given below due to its importance and relevance to this work and is as follows:

> "For the purposes of delimitating the Cotteswold Hills, the 400 foot contourline was found to be in most accord with the extent of the area under consideration. On the western side it is certainly the best line to take: a lower one would include too much, a higher one would define too little. Only in one case, in the north-east part, is it necessary to rise above the 400 foot line; still, on the whole, it may be said that what is above the 400 foot line in the area to be described is Cotteswold country.
>
> Starting from Bath, we can follow the 400 foot line around Lansdown Hill and proceed northwards to Sodbury, Hawkesbury, Stinchcome Hill, across the Dursley Valley to Cam Long Down, and thence by Selsley to the Stroud Valley. Here we must take an arbitrary line, circling from one hill to another around the town of Stroud, then proceeding again on the 400 foot line from Randwick Hill. This takes us easily enough to the northern part of the hills; and as the northern-most part of the Cotteswold range we must place Meon Hill, which is connected by a narrow neck above the 400 foot level, to the high ground of Ilmington, Hidcote Bartrim, and Ebrington. The 400 foot line can then be followed till we come to the watershed between the Knee Brook and the Evenlode. Here it is necessary to take an arbitrary line to the head of the Evenlode and proceed down the right hand bank of that river on the 400 foot contour again. This line, of course, runs about the middle of the Vale of Moreton. But that vale must be regarded as a high level vale - much of it is above 400 feet - parting the Cotteswolds from the high ground around Chipping Norton and Rollright, which seems to have no name. The 500 foot contour line from Aston Magna to Fifield, would be about the western delimitation of the Vale of Moreton, and the Vale must be considered to belong partly to the Cotteswolds in a sense, and partly to the hills of Oxfordshire.
>
> We now follow along the Evenlode, around Wychwood Forest, to Wilcote.
>
> It does not seem to part the high ground of Wychwood and Leafield from the Cotteswold range. It is connected with the ridge of Barrington and Rissington by ground over 500 feet high; and as Burford with the country round it, is considered Cotteswold, the best physical boundary on the north-east side is the Evenlode: that of necessity makes the area of Wychwood Forest an eastern extension of the Cotteswolds.
>
> It would be difficult to take the Windrush as the boundary as has been suggested. That would cut off the Barrington and Rissington country. And if the Windrush were followed further up it would cut off most of the North Cotteswolds-the very area of the classic Cotteswold games. It certainly seems that the valleys of the Stour and the Evenlode form the best physical parting between the Cotteswolds and the high ground of Oxfordshire and Warwickshire.
>
> For the south-eastern boundary from Witney I have taken roughly the limit of the Cornbrash, where it dips under the Oxford Clay. This forms a slight feature, it enables us to include in the Cotteswold area the typical arable stonewall country; it runs between the 300 and 400 foot contourlines, though there is some variation. By this line we include in Cotteswolds, Brize Norton, Broughton, Fairford, Siddington, Crudwell.
>
> This line I keep as near to Easton Grey. There the Cornbrash area is so merged into the grassland of the Vale of Chippenham, that it cannot be taken as the limit. So I have taken two rivers, the

> Sherston branch of the Avon up nearly to Acton Turville, then an arbitrary line to Burton, and thence down the By Brook to Bath."

Conybeare and Phillips (1822, p. 220-225) suggested that

> "The Evenlode seems to form the natural boundary between the system of hills connected with the uplands of North Oxfordshire and those dependent on the Cotteswold Hills..."

Richardson (1935) and Richardson *et al.* (1946) tended to disgree with Buckman's scheme for the north eastern Cotswolds and Richardson's inclination was that this boundary should follow the bottom of the 'Maugersbury Gap' between Stow-on-the-Wold and Church Icomb to the River Windrush at Bourton-on-the-Water, the Windrush down the Vale of Bourton to the Oxfordshire county-boundary near Burford, and then along the county-boundary southward to the outcrop of the Cornbrash (Beckinsale, 1992, p.24-25). Despite his challenge to Buckman's scheme, Richardson offered no real geographical or geological explanations for his preferences.

Beckinsale (1934) undertook a detailed investigation into the boundaries of the Cotswolds and strongly approved the thoughts of Buckman, with only a few minor revisions to his map. In this account the delimitation of the Cotswolds shall follow this scheme (*sensu* Beckinsale, 1934, 1992), the outline of which is shown in Figure 1.

Traditionally, the Cotswolds are said to terminate northwards of Kingham and Long Hanborough, but the upland immediately north of the Evenlode gorge is capped with Great Oolite limestone and its landscape is characteristically Cotswold. This area forms a dissected plateau near Chipping Norton and the Cherwell. This area is known as the Chipping Norton Heights with Edge Hill to the north of this area. The scarp of Edge Hill rises to over 212 mOD, and marks the north-western limit of the Middle Lias plateau. The flat summits lack height when compared with those of the Cotswolds and are never far from deep river valleys, which in some places have incised into the Lower Lias. One such stream completely separates the Burton Hills-Farnborough-Burton ridge from the main plateau. This long ridge is 211 mOD high at Church Hill, Burton Dasset, and declines gently south eastwards to 127 mOD at Little Bourton about three kilometres north of Banbury.

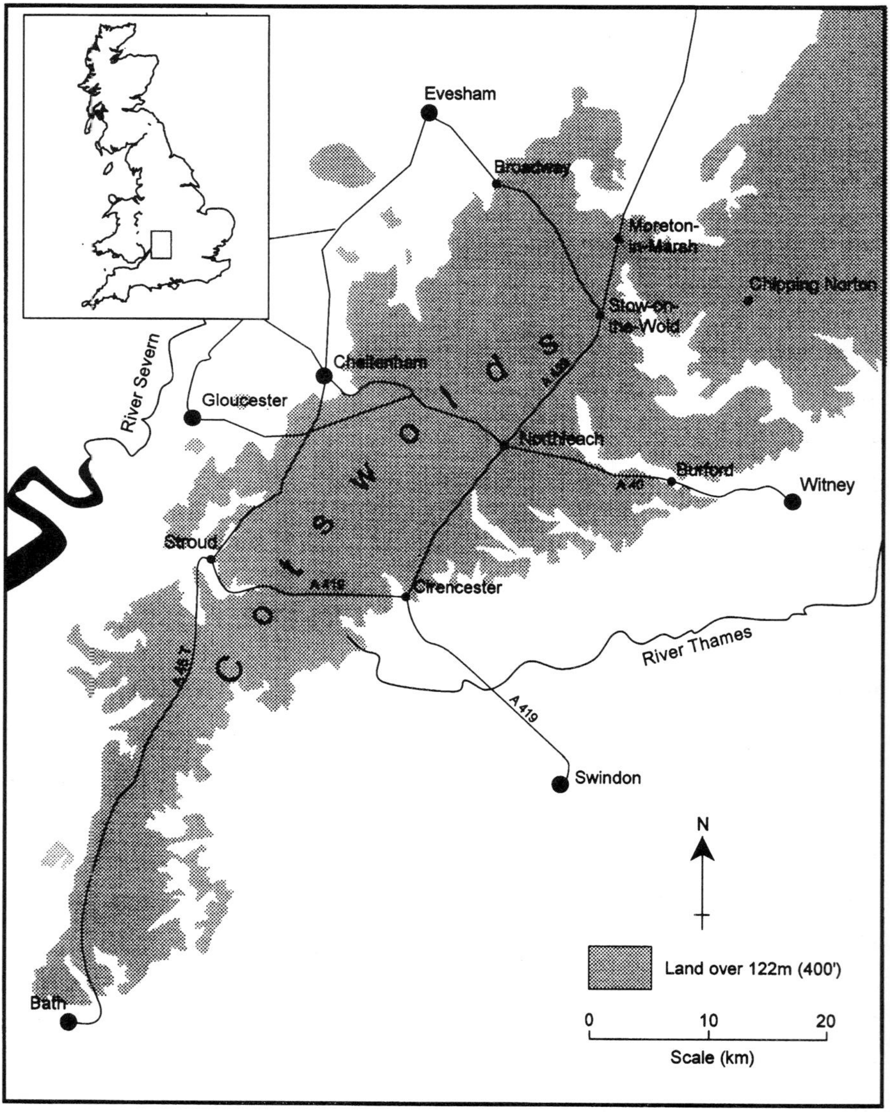

Figure 1 The location of the Cotswolds

Chapter 2

2.1 Geology: bedrock

The strata within the Cotswolds were deposited during the Mesozoic and Cainozoic Eras. The geology comprises Jurassic rocks which crop out along the backbone of England from Dorset and run north to Yorkshire. These sheets of Jurassic sediments are lithologically disparate, with their present day disposition determined by tectonic activity during the Cretaceous and subsequently. Such tilting and folding produced a regional dip of a few degrees to the south-east in this area and therefore they appear at the surface as bands of strata striking south-west to north-east (Martin and Steel, 1954). These deposits are sedimentary in origin and consist mainly of marine clays, sands and limestones. The Jurassic rocks, including clays, attain a maximum height of 330mOD (1082') at Cleeve Common, 313mOD (1025') at Broadway Hill and 319mOD (1046') above Seven Springs. Heights of around 250mOD are attained on the northern boundaries (e.g. at Rollright), and on the eastern side heights of 90 mOD occur around Little Faringdon.

While the general geological structure is indeed relatively simple, with broad beds of essentially N-E to S-W striking limestone and clay formations, this simplicity hides a wealth of local detail, with differentiation in the overall simple pattern of relief. Arkell (1947a, p.11) pointed out that the stratigraphical geology is

> "...primarily a study of wedge-shaped and incomplete formations full of eroded surfaces and disconformities, often only detected through palaeontological non-sequence.. "

Gaps in the record, the rhythmic alternation of deposits, the effects of flexing and faulting, changes in thickness and local dip of the strata, give rise to an exceedingly complex distribution of beds and a great deal of uncertainty concerning detailed correlation.

Figure 2 shows a generalised map of the solid geology of the Cotswolds, whilst Table 1 gives the order of the beds encountered. The first published geological book specifically on the area was Murchison's (1834) *An Outline of the Geology of Cheltenham*, which was followed by a larger second edition revised by Strickland and Buckman (1844).

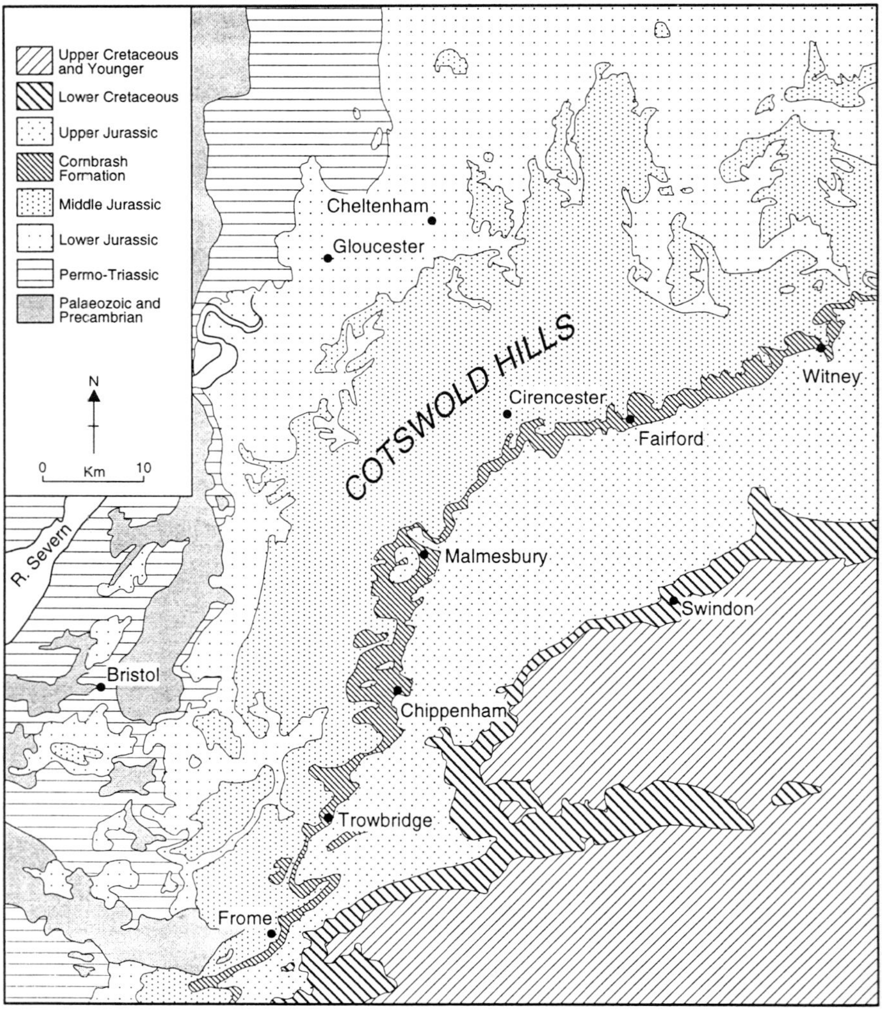

Figure 2 Major geologic units in the area (after Chidlaw and Campbell, 1989, fig1)

STRATIGRAPHICAL DIVISION	STAGE	FORMATION
Oxford Clay	Lower Oxfordian	Upper Oxford Clay
	Callovian	Middle Oxford Clay Lower Oxford Clay
Kellaway Beds		Kellaways Rock Kellaways Clay
Cornbrash		Upper Cornbrash
	Callovian Transgression	
	Upper Bathonian	Lower Cornbrash Wychwood Beds
Forest Marble		Bradford Beds Kemble Beds
Great Oolite Series	Middle Bathonian	White Limestone Hampen Marly Formation Taynton Limestone Formation Sharp's Hill Formation Acuminate Beds
Chipping Norton Limestone / Fuller's Earth Clay	Lower Bathonian	Chipping Norton Limestone / Fuller's Earth Formation
Inferior Oolite Series	Upper Bajocian	*Clypeus* Grit Upper *Trigonia* Grit
	Arkell's (1933) Vesulian Transgression	
	Middle Bajocian (Bajocian s.s.)	*Sauzei* Zone *Witchellia* Grit Notgrove Freestone *Gryphite* Grit *Buckmanii* Grit Lower *Trigonia* Grit
	Arkell's (1933) Bajocian Transgression	
	Lower Bajocian (Aalenian)	Tilestone Snowshill Clay Harford Sands Overlap: break in succession Upper Freestone Oolite Marl Lower Freestone Pea Grit Lower Limestones *Scissum* Beds
Cotteswold Sands	Yeovilian	Cotteswold Sand
Upper Lias Clay	Whitbian	Upper Lias
Middle Lias Clay	Pleinsbachian	Middle Lias
Lower Lias Clay		Lower Lias

Table 1 Stratigraphy and lithological characteristics of the Jurassic rocks of the north Cotswolds (after Hart, 1976)

Subsequent detailed accounts of the regional geology may be found in Phillips (1871), Richardson (1904a), Arkell (1947a), Martin and Steel (1954), McKerrow and Kennedy (1973), Dreghorn (1967) and Green (1992) and the solid and drift one inch, 1:50 000 and 1:10 000 British Geological Survey's geological maps and memoirs for this area. These memoirs include: Edmonds *et al.* (1965), Banbury; Horton *et al.* (1987), Chipping Norton; Richardson (1929), Moreton-in-Marsh; Richardson (1933), Cirencester; Richardson *et al.* (1946), Witney; Cave (1977), Malmesbury; Hull (1857), Cheltenham; Green (1992), Bristol and Gloucester; Worssam *et al.* (1989), Tewkesbury and Williams and Whittaker (1974), Stratford-upon-Avon and Evesham.

Beneath the Jurassic rocks are the clays, marls and sandstones of the Trias. They are not exposed at outcrop within the designated area of the Cotswolds but are known from borehole evidence to be present at depth.

The geological sequences found within the Cotswolds shall now be discussed in ascending order, as they are of fundamental importance when considering the geomorphology of the region and local variations within it.

2.2 The Lower Jurassic System

2.2.1 Lower Lias

The oldest rocks which appear at the surface in the Cotswolds are Lower Lias Clays, of Lower Jurassic age. The name 'Lias' was given to the Series by William Smith, though the term had been used as far back as the 17th century (Richardson, 1904a, p.35). The Lower Lias underlies almost the whole of the region, but only appears at the surface in the north of this area along the valley floors of the Cherwell and Evenlode (Arkell, 1947a,b; Martin and Steel, 1954). It also floors the initial portions of the Dursley, Stroud and Chelt Valleys, the Valley at Winchcombe and almost completely the enclosed Vale of Bourton. The Lower Lias Clays form the lowland in the Severn Vale at the foot of the Cotswold escarpment. Their depth is uncertain, though at Cheltenham it has been estimated to be between 150-300m.

The Stowell Park Borehole between Northleach and Cirencester has yielded very important information for the Lower Lias at its maximum development in the Cotswolds as shown in

Figure 3. When traced westwards the Lower Lias shows a marked decrease in thickness, from 140m to 65m (Green and Melville, 1956).

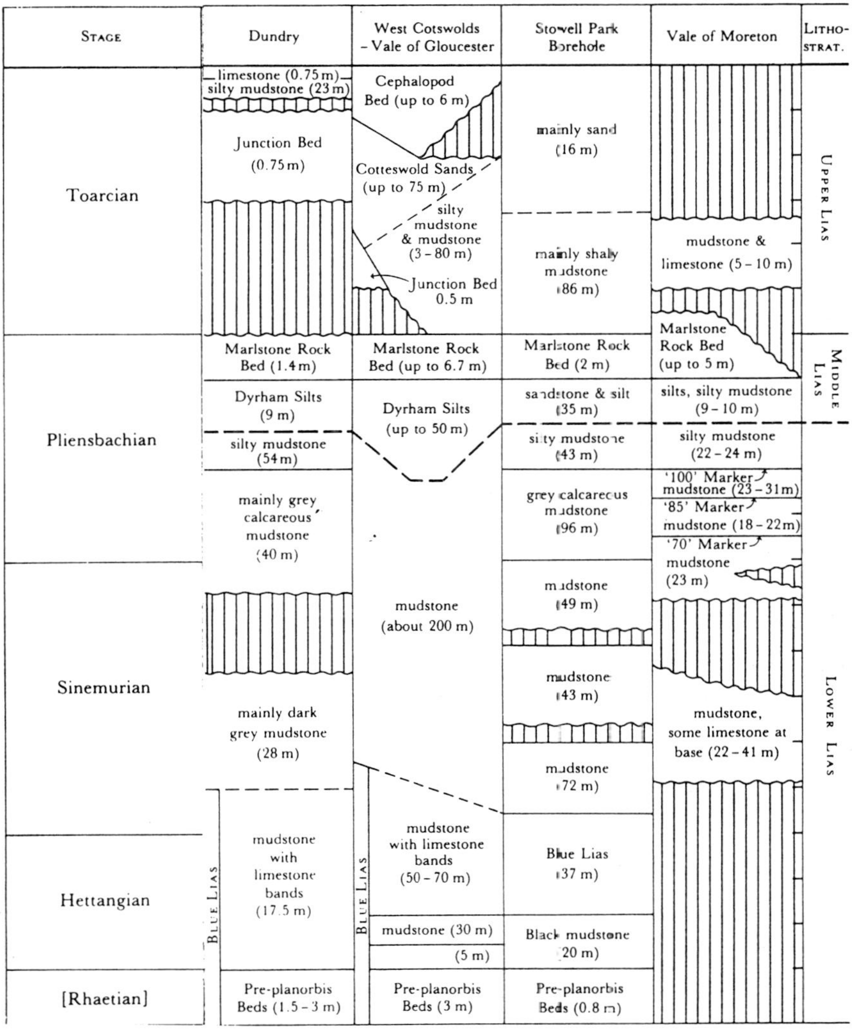

Figure 3 Lithostratigraphy and chronostratigraphy of the Lias formations (after Green, 1992)

In general the Lower Lias comprises somewhat sandy clays with associated beds of impure limestone, passing down into the blue clay which constitutes the main mass. Upon weathering the medium grey mudstones change to pale-bluish grey and brown mottled clay. The highest layers of the Lower Lias pass into highly calcareous marly bands such as the Belemnite Marls (Findlay, 1976). One of the best sections in the Lias in England is to be found at Tuffley Old Brickworks at Robins Wood Hill (Ager *et al.*, 1973), which is an outlier of the Cotswold escarpment.

2.2.2 Middle Lias

The Middle Lias comprises two formations, a lower, thick silty and arenaceous group (the Dyrham Silts) and an upper, much thinner, ferruginous shallow-water limestone (the Marlstone Rock Bed) (Fig. 3). The Middle Lias represents a regressive phase in sedimentation. Middle Lias material is absent from the extreme southern portion of the south Cotswolds, where Upper Lias rests directly on Lower Lias.

The Dyrham Silts (formerly the Middle Lias Silts) are at their typesite (Dyrham) about 20m thick. They are relatively thin in south Gloucestershire. Northwards they increase to 75m in the mid-Cotswolds. In the north Cotswolds-Chipping Norton Heights-Edgehill area the maximum outcrop thickness may exceed 45m. This thins to the south-east, but superimposed on this regional trend are local variations. The beds are pale to mid grey micaceous silts and silty clays. Ferruginous nodules occur throughout along with shell beds and lenses. The Dyrham Silts form a steep slope beneath the Marlstone Rock bed, often obscured by slipped Upper Lias Clay. The junction between the Middle Lias and Lower Lias is marked by a change in slope, with common seepage lines usually at the base of the more permeable Dyrham Silts. The Dyrham Silts are abruptly succeeded by the Marlstone Rock Bed.

The Marlstone Rock Bed comprises a hard massive limestone, blue- or green-coloured when fresh but readily weathered to a rich brown and reddish brown hue. It forms a broad elevated plateau between the Warwickshire border and the River Swere. This extends northwards into the Banbury District to culminate in the Edge Hill escarpment (Edmonds *et al.*, 1965). The valleys tend to be steep sided and are cut into the Lower Lias Clay. The valley of the River Swere is an exception to this. This is a graben, in which the Oolitic limestones are preserved at depth. The Marlstone Rock Bed outcrops are extensively cambered (see section 5.3 below), and thickness estimates are difficult (Horton *et al.*, 1987).

The plateau has outliers of Upper Lias on it to the west and south of Banbury, and also runs along the east side of the River Cherwell to Rousham, with a smaller fragmented strip also occurring on the western side. Narrow strips also crop out on both sides along the Evenlode valley. The plateau of the Middle Lias is comprised of a sandy limestone coloured with iron oxides, which has been quarried for iron ore around Banbury.

The Marlstone Rock Bed forms a conspicuous ledge below the Cotswold scarp, a feature which is well exhibited near Wotton-under-Edge and Stinchcombe. It also gives rise to the flat-topped hills of Meon, Dixton and Dumbleton, is found in the outliers of Ebrington, Nebsworth, Dover's, Dundry and Bitton Hills, and the ledges on Oxenton, Alderton and Bredon Hills, between Cheltenham and Evesham.

2.2.3 Upper Lias

Above the Marlstone Rock Bed lie Upper Lias deposits. These consist predominantly of medium to pale grey mudstones, which weather to pale greyish brown clays and overlie a thin, dominantly calcareous unit that crops out on the gentle slopes above the Marlstone Rock Bed. The maximum total thickness of the Upper Lias in the North Cotswolds is *ca.* 60m at Sharp's Hill but pinches out in the south east. In the mid Cotswolds the beds are around 90m in thickness. Figure 4 shows the Upper Lias for the Cotswolds (after Kellaway and Welch, 1948). The outcrop, which is poorly drained, tends to be narrow, protected by the overlying Middle Jurassic limestones. The Upper Lias sediments display markedly different sedimentation patterns in the north and south Cotswolds. In the south the sediments are thin, but thicken in the mid and north Cotswolds (see Figure 4).

The Upper Lias is divided, in ascending order, into the Transition or Junction Bed, the Fish Beds Member, the *Cephalopod* Limestone Member and an unnamed argillaceous unit (Horton *et al.*, 1987).

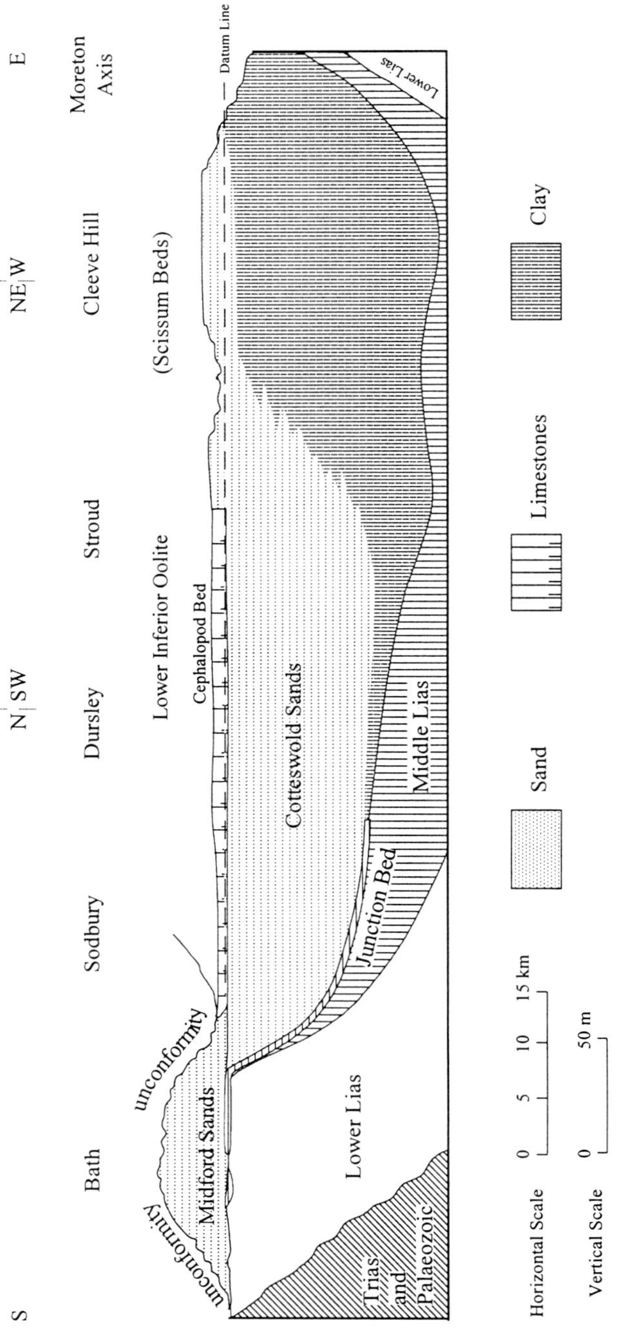

Figure 4 Diagrammatic section of the Upper Lias to illustrate facies changes across the region (after Kellaway and Welch, 1948)

The Junction Bed consists of thinly bedded grey limestone with ferruginous, oolitic and conglomeratic beds. At Old Sodbury this unit is represented by 1.5m of cream-coloured marl, limestone and conglomerate. North from here the Junction Bed passes into sands and clays and ceases to be a recognisable feature. In the north Cotswolds the sediment is referred to as the Transition Bed (Walford, 1878; Thompson, 1910; Horton *et al.*, 1987), represented by 0.13m of silty limestone, as shown in the Steeple Aston Borehole.

Much of the Upper Lias comprises silts and fine sands of the Cotteswold Sands. These are often associated with slope instability and cambering. Around Winchcombe, they attain a maximum thickness of *ca.* 85m. At Cleeve Hill they are *ca.* 90m in thickness. The Cotteswold Sands thin rapidly north of Stroud and give way to the Upper Lias Clay. The *Cephalopod* Bed attains a maximum thickness of 4.6m in the Dursley district, where it consists of ferruginous, oolitic limestones and marls with abundant ammonites and belemnites.

In the southern Cotswolds, the basal layers become sandy and then thicken as they pass into the Midford Sands which in turn pass in the north into the Cotteswold Sands. Judd (1875) formerly described these deposits as the Sepentine Beds, Thompson (1910) as the Upper *Cephalopod* Clay and Lower *Cephalopod* Clay and Horton *et al.* (1974) as the *Cephalopod* Limestone Member. The member comprises brownish-grey weathering, fine grained, slightly oolitic limestones with slightly darker grey marly clays and mudstones.

An unnamed unit overlying the *Cephalopod* Bed has been described by Horton *et al.* (1987), which was known as the Unfossiliferous Beds of Richardson (1923a). It comprises medium grey mudstones with scattered phosphatic and ferruginous nodules, and contains very few macrofossils.

The whole of the Middle and Upper Lias succession generally outcrops on moderate to steep slopes of the Cotswold scarp where its form is invariably complicated by spring sapping, cambering and slipping.

2.3 The Middle Jurassic System

The stratigraphy of the Middle Jurassic, the rocks of which give rise to to the Cotswold cuesta, have received considerable attention from geologists. They comprise a varied sequence of interbedded marine and lagoonal sediments, many separated by minor non-sequences. Despite the wealth of information available, the Middle Jurassic system remains controversial. The term 'Series' has been replaced by 'Group' (Horton *et al.*, 1987). The Middle Jurassic may be divided into the Inferior and Great Oolites, and will now be discussed.

2.3.1 Inferior Oolite Group

The Upper Lias is succeeded by a sequence of shallow water marine limestones known as the Inferior Oolite Group, first identified by William Smith. Smith used the terms 'Under Oolite' or 'Inferior Oolite', and the latter term was extended by Townsend (1813). This Group extends through the Aehenian, Bajocian and Bathonian Stages, and near Cheltenham the Inferior Oolite attains a thickness of *ca.* 100m. It constitutes the summits of the main Cotswold mass and forms the great indented scarp of the Cotswolds. S.S. Buckman, at the turn of the century, published a series of important papers on the Middle Jurassic of the Cotswolds (Buckman, 1895, 1897) in which he sub-divided the Middle Inferior Oolite and defined the relationship of the lower, middle and upper parts to each other. He realised that the previous research had failed to recognise the complex stratigraphy. His key to understanding this complex situation was to construct a stratigraphy chiefly based on the ammonite biostratigraphy, and where these were rare, by the use of brachiopods. His research was initially misunderstood and received much hostility, but despite this early problem his stategy is widely used today. The classification of the Inferior Oolite Group is shown in Table 2 and Figure 5. The thickest succession of Middle Jurassic Rocks is at Cleeve Hill, near Cheltenham, and the sequence is shown in Figure 6.

The lower Inferior Oolite was re-examined by Mudge (1978), in which a new nomenclature was ascribed, though this has been challenged by Cope *et al.* (1980).

2.3.1.1 *Scissum* Beds

The lowest group are the *Scissum* Beds which comprise yellow to brown, fine-grained, sandy limestones and calcareous sandstones. The rocks may have a somewhat sparse fauna. In the north Cotswolds the formation consists of yellowish grey and pale-orange weathering sandy limestones (Richardson, 1911), which are known as the Northampton Sands.

Variations in thickness have been suggested as being due to pre-*Clypeus* Grit erosion, rather than to sedimentation differentiation across the region (Horton *et al.*, 1987). The beds usually rest unconformably on the Upper Lias, the base of which can be difficult to locate where they directly overlie Cotteswold Sands.

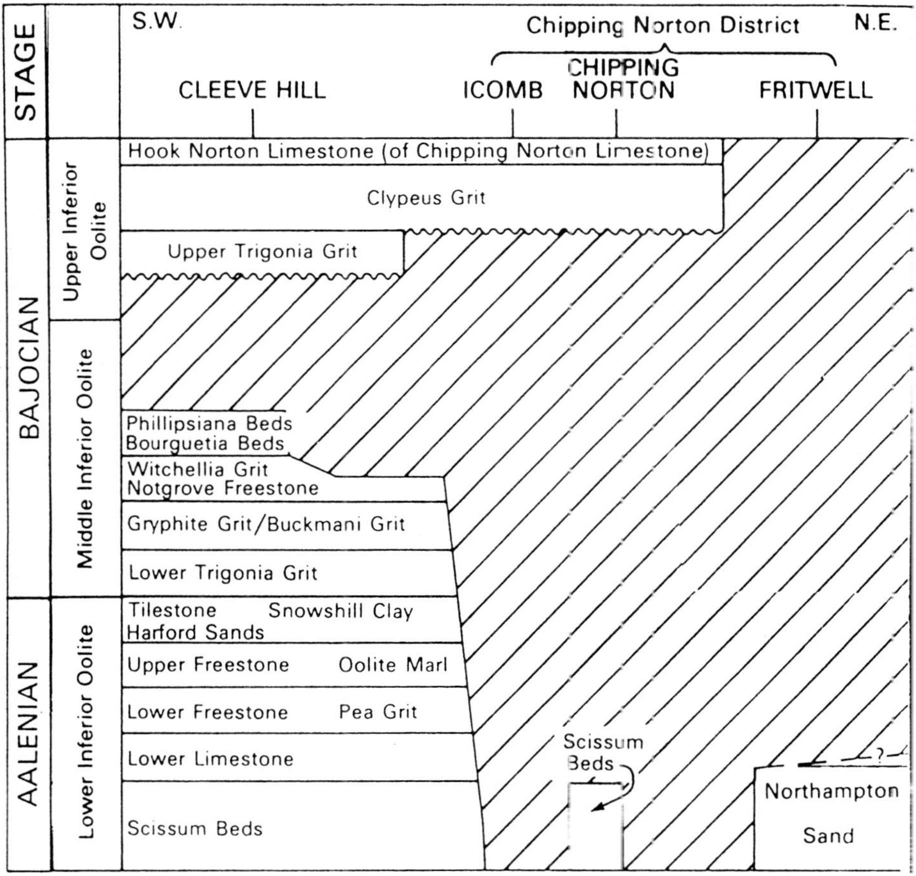

Table 2 Classification of the Inferior Oolite Group (after Horton et al., 1987)

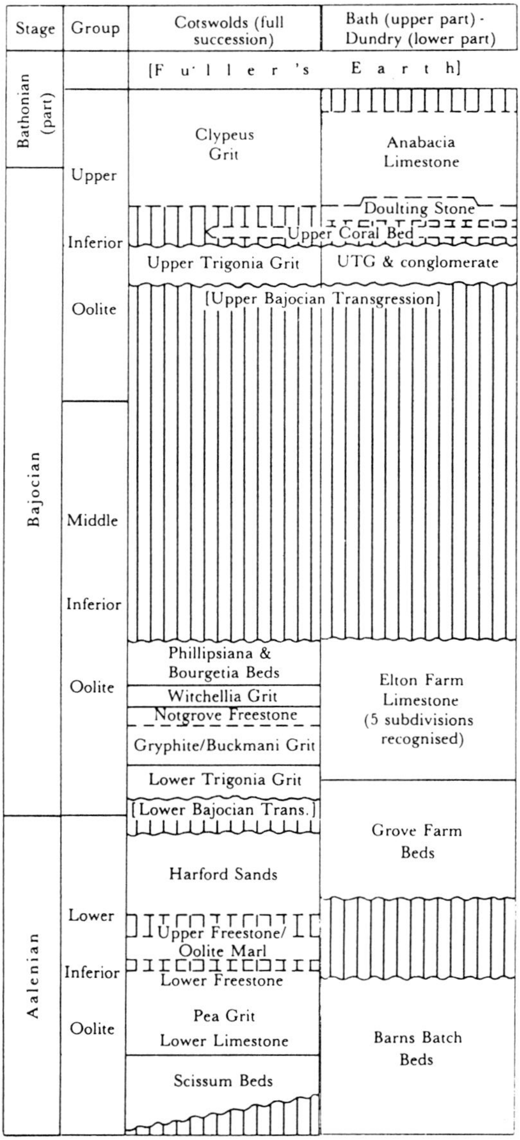

Figure 5 Classification of the Inferior Oolite Group (after Green, 1992)

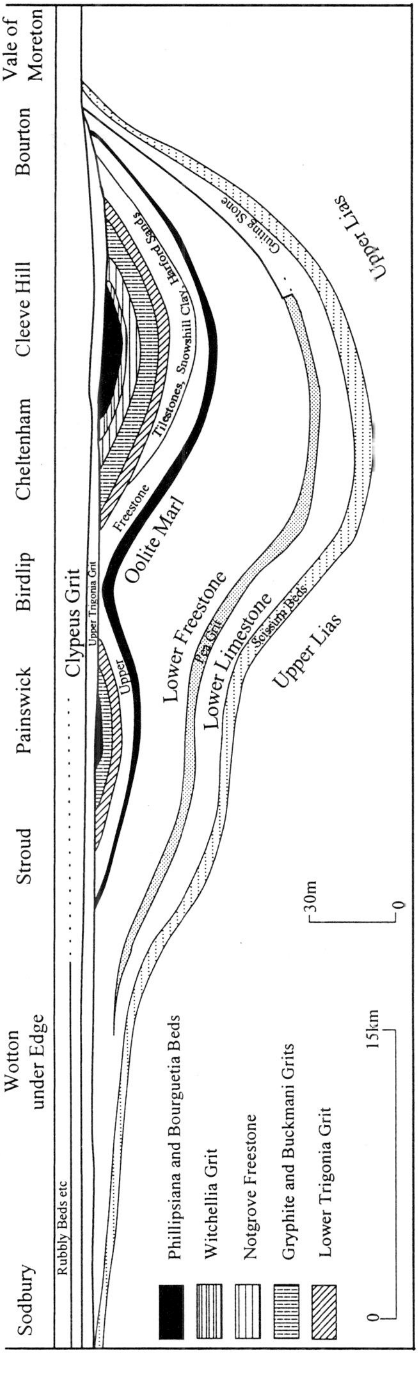

Figure 6 Diagrammatic cross-scetion through the Inferior Oolite of the Cotswolds (after Arkell, 1947a)

Witchell (1886) observed this deposit 'is persistent throughout the greater part of the Cotteswolds' and named it the 'sandy ferruginous limestones'. The beds were laid down during what Richardson (1904a) describes as the hemera (sub-stage) *Scissi*, the characteristic ammonite, *Tmetoceras scissum*, being the most abundant type found. Overall the thickness is rarely more than a few metres. Richardson (1904a) reports 2.25m of *Scissum* Beds at Haresfield and 4m at Coopers Hill near Andoversford. Between Stanway Hill and Syreford there are no definite sections. The outliers of Crane and Oxenton Hills are capped by *Scissum* Beds, though they are very disturbed. At Bredon Hill, these beds comprise brown arenaceous strata, which are very ferruginous in places.

2.3.1.2 Lower Limestone

The Lower Limestone in general comprises up to 10m of unfossiliferous, massive current-bedded oolite, which rests on the eroded surface of the *Scissum* Beds. The beds vary considerably across the Cotswolds and in places the division is arbitrary e.g. at Crickley Hill (Richardson, 1904a). Here, Witchell (1886) recorded 6m of the Lower Limestone. On Bredon Hill the Lower Limestone Beds are thicker, *ca.* 10m, and contain numerous fragments of echinoid spines (Richardson, 1904a),

In the Stroud District the beds contain coarse sand, and locally rounded grey quartzite pebbles. In the north-east Cotswolds and passing in to the Chipping Norton-Edgehill district, the Northampton Sands are overlain by 3-6m of pale grey and brown sands known as the Lower Estuarine Series (Horton *et al.*, 1987). They were originally grouped with the Northampton Sands (Aveline and Trench, 1860) and were thus considered part of the Lower Inferior Oolite, a view supported by Horton (1977) and Horton *et al.* (1987). Other workers have viewed them as much younger, equating to the Swerford Beds, an arenaceous facies of the Chipping Norton Limestone (Richardson, 1911; Arkell, 1947a; Sellwood and McKerrow, 1974).

2.3.1.3 Pea Grit

The Pea Grit is scarcely represented in the south Cotswolds. It is present in the central area around Crickley Hill-Leckhampton Hill-Cleeve Hill. The Pea grit comprises a variable group of rubbly, pelloidal and sparsely pisolitic limestone with interbedded oolitic rocks. It attains a thickness up to 12m.

2.3.1.4 Lower Freestone

The Lower Freestones rests on the Pea Grit. It comprises a fine-grained, light coloured oolite, which closely resembles Bath Freestone (Brodie, 1850; Richardson, 1904a), and is one of the thickest and most distinctive members of the Lower Inferior Oolite. It has been extensively quarried and used for building stone, especially in Cheltenham. On Cleeve Hill, borehole evidence has revealed over 50m of Lower Freestone.

2.3.1.5 Oolite Marl-Upper Freestone

Buckman (1886) suggested that the two were one and the same, and recent work by Baker (1981) supports this view. Indeed, workers using the former bi-partite system considered the two together as in many places it is difficult to separate them, for example in the district to the north of Stroud. The beds are highly distinctive and rest nonsequentially on the Lower Freestone. Three main facies are present and comprise marls, calcite-mudstones and limestones.

2.3.1.6 Harford Sands, Snowshill Clay and Tilestone

These form the highest beds of the Lower Inferior Oolite, and are restricted to the region between Winchcombe and Broad Campden in the north Cotswolds. They occupy a core of the Cleeve Hill Syncline where they attain a maximum thickness of 13m (Figure 6). The Harford Sands consists of 3m of pale brown quartz sands which contain hardened sands called doggers (Richardson, 1904a; Green, 1992). The Snowshill Clay is a chocolate-coloured clay, 4-5m thick, which is at its maximum thickness at Blockley. The overlying Tilestone consists of sandy, oolitic, flaggy limestone.

2.3.2 The Middle Inferior Oolite Group

The Middle Inferior Oolite group is only preserved in two synclinal areas (Figure 6). They comprise a mix of thin-bedded sediment facies which include the Lower *Trigonia* Grit, *Buckmanii* and *Gryphite* Grits, Notgrove Freestone, *Witchellia* Grit and *Phillipsiana* and *Bourguetia* Beds. Detailed accounts of these minor beds can be found in Richardson (1904a) and Green (1992).

2.3.3 The Upper Inferior Oolite Group

The *Clypeus* Grit is the sole representative of the Upper Inferior Oolite in the region. It comprises pale, coarse-grained, shelly oolitic micritic limestone. In the north Cotswolds a non-sequence separates the Scissum Beds from the overlying *Clypeus* Grits. At Shilton, borehole evidence has revealed a thickness of 23m. To the west this thins to 2m around Fawler in the valley of the Evenlode. It is not frequently seen on the escarpment, though it is found at Birdlip (Richardson, 1904a). Richardson has also recorded sections at Seven Springs, Cleeve Hill, Hampen, Andoversford, Compton Abdale, Yanworth and Taynton.

2.3.4 Great Oolite Group

The Middle Jurassic succession has traditionally been grouped as the Great Oolite Series, but this has now been replaced by the term Great Oolite Group. This group embraces all the formations between the Inferior Oolite and the Kellaway Clays. The sequence is largely of Bathonian age. The Great Oolite Group comprises a large number of limestones, clays and grits, which were examined and discussed in detail by Arkell and Donovan (1951). The total thickness is between 60 and 100m. The succession displayed in the stratigraphical record has broadly been interpreted as resulting from fluctuating marine conditions, with periodic brackish and freshwater episodes. As for the Inferior Oolite group, nowhere are all the formations present in one oucrop in the Cotswolds. The stratigraphy of the Great Oolite Group for the north Cotswolds is shown in Figure 7.

2.3.4.1 Chipping Norton Limestone

The Inferior Oolite Group is succeeded in the north Cotswolds by the sandy, oolitic Chipping Norton Limestone. The Chipping Norton Limestone was named by Huddleston (1878) and covers the strata between the Clypeus Grit and the Sharp's Hill Bed. It attains a maximum thickness at Chipping Norton, where it is 12.7m thick. To the west and south west of the north Cotswolds it passes into Fullers's Earth. Typically it comprises buff, hard, rather sandy and splintery oolite. The beds are in places flaggy, whilst in other they have been decalcified and reduced to sand (Horton *et al.*, 1987). The late Marjorie Sweeting (pers.comm) believed this was the most karstified part of the Oolite.

2.3.4.2 Fuller's Earth

The Fuller's Earth Formation, which is an important control of landslide activity, is an overconsolidated, jointed, randomly fissured, calcareous mudstone with some thin to medium, jointed argillaceous limestone bands (Hawkins *et al.*, 1988).

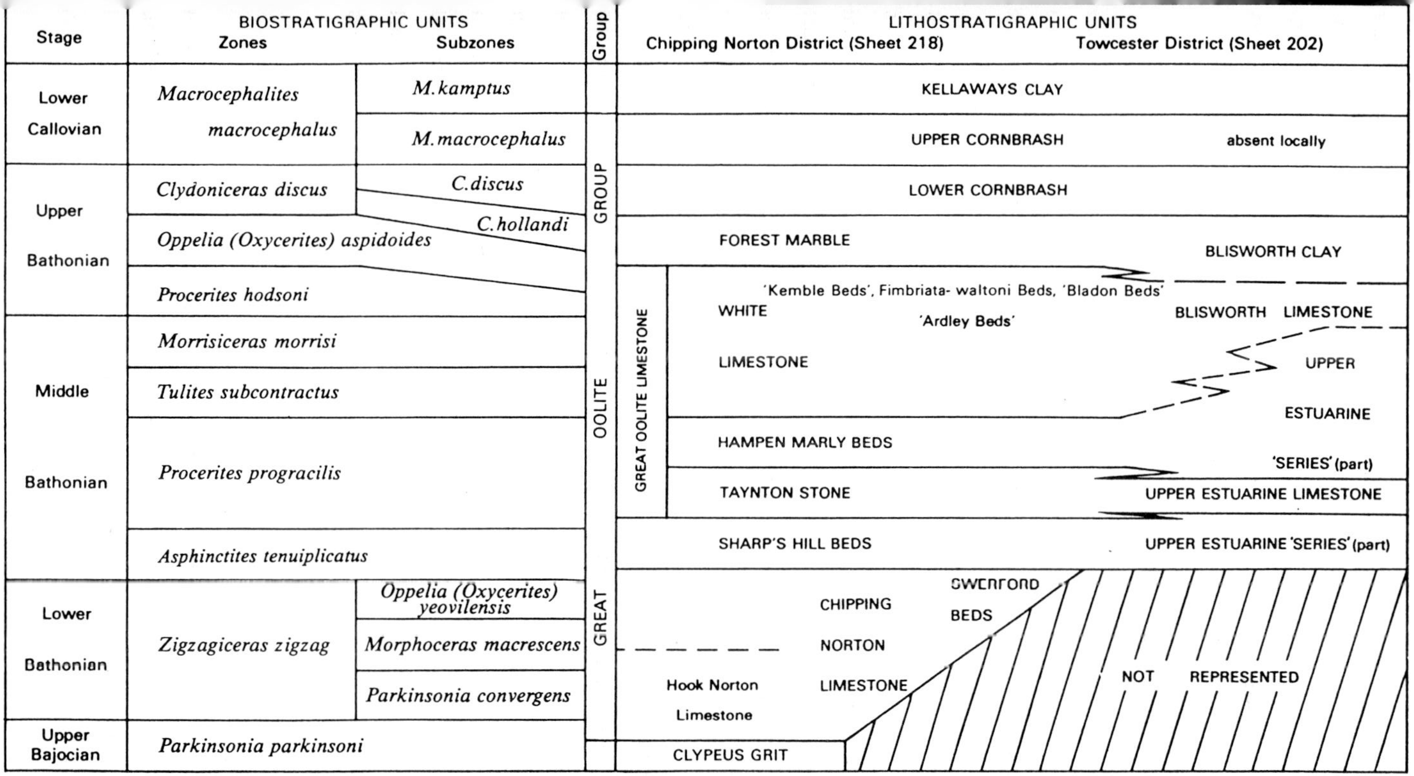

Figure 7 Stratigraphy of the Great Oolite Group (after Horton et al., 1987)

Bath is the type site and the stratigraphic sequence has been described by Kellaway and Welch (1948) and Arkell and Donovan (1951). The sequence is as follows: Lower Fuller's Earth *ca.* 10m, Fuller's Earth Rock *ca.* 4m, Fuller's Earth Bed *ca.* 2-3m and the Upper Fuller's Earth *ca.* 25m. Northwards from Bath, the formation becomes increasingly calcareous, and several of the argillaceous limestone bands observed in the southern Cotswolds coalesce to develop thick, individually named units (Hawkins *et al.*, 1988).

In the north Cotswolds, the Lower Fuller's Earth thins rapidly and passes into the Stonesfield Slate Beds between Minchinhampton and Cirencester. It is absent north of Condicote and east of the Moreton 'Axis' (Green, 1992), where it is replaced by the Chipping Norton Limestone and Sharp's Hill Formation. The Upper Fuller's Earth is absent in the north Cotswolds.

2.3.4.3 Sharp's Hill Formation

The Chipping Norton Limestone is conformably succeeded and overlain by the Sharp's Hill Formation which comprise bright coloured and grey, highly fossilerous clays and thin shelly limestones. Walford (1906) formerly called them the Naeran Beds, but after several revisions the term Sharp's Hill Bed (now superceded by Sharp's Hill Formation after McKerrow and Kennedy, 1973) was applied to the fossiliferous limestones, muds and clays between the Chipping Norton Limestone and the Taynton Stone. In the west of the area it now includes the newly recognised Charlbury Formation (Boneham and Wyatt, 1993). The Sharp's Hill Formation rests conformably on the Chipping Norton Limestone, except in the very east of the region, where they overlap the Lower Estuarine Series.

2.3.4.4 Stonesfield Slate Beds

These beds comprise 4-10m of fissile, sandy, shelly and oolitic limestones, which were formerly worked for roofing slates. Their precise age is uncertain, with certain workers placing them either below or above the Sharp's Hill Beds (Arkell, 1947a; Selwood and McKerrow, 1973).

2.3.4.5 Great Oolite Limestone

Bath is the type area for the Great Oolite, which has been described by Green and Donovan (1969). The succession is 20-30m thick and continues from Bath to Castle Combe. It comprises the Combe Down Oolite, a cream-coloured, cross-bedded oolite freestone with shell detrital marly oolitic limestone. The Twinhoe Beds are a buff, fine-grained, shell detrital limestone with some rubbly intercalated limestones. The dominant facies of the Great Oolite Limesone is the Bath Oolite, a cream coloured, cross-bedded oolite freestone.

In the north Cotwolds the Great Oolite Limestone comprises in upward succession, the Taynton Limestone, the Hampen Marly Beds and the White Limestone. The Taynton Limestone Formation comprises cream and buff, coarse-grained, shell-detrital oolitic limestones, generall cross-bedded, flaggy or massive which is between 3-12m in thickness (McKerrow and Kennedy, 1973; Horton *et al.*, 1987). It has been extensively quarried in the past as a freestone for building purposes.

The Hampen Marly Beds (now Hampen Formation (Sumbler and Barron, 1996, in press)) which overlie the Taynton Limestone Formation are poorly exposed. Arkell (1931) first decribed this formation from Hampen, near Cheltenham. This ranges between 4-13m in thickness, and comprise limestones and marls with thin mudstone, siltstone and sandstone bands. In the north of the Cotswolds and the Banbury-Edge Hill area the Hampen Formation passes into the Upper Estuarine Series. The Marly Beds locally produce acidic soils and are zones of seepage

The White Limestone comprises limestones interbedded with clay beds. The limestones are generally weak to moderately strong. The rock is well jointed and may be frost susceptible. Sumbler (1984) has discussed the subdivisions of the White Limestone in detail.

2.3.4.6 Forest Marble

The Forest Marble, named by William Smith, is a hard, flaggy, false-bedded, broken shell limestone, oolitic, between two argillaceous divisions. These clays are sandy and micaceous, with numerous thin impersistant laminae of hard, fissile or broken shell limestone. In general the limestone bands are strong but the finer grained material is often frost susceptible. The clays, which are typically unbedded, are grey or green when fresh and weather to whitish and brown respectively. In the area around Burford the Forest Marble is around 5-8m thick. Around Cirencester the limestone component is variable but well developed. In the Malmesbury-Kemble area 9-14m of these beds have been recorded. In the southern Cotswolds between Bath and Malmesbury the Forest Marble thickens to between 26 and 28m (Cave, 1977).

2.3.4.7 Cornbrash Limestone Formation

The final rock at outcrop to be considered as 'Cotswold' in the sequence is the Cornbrash Limestone Formation which rests with marked disconformity over the Forest Marble, and is the uppermost formation comprising the Great Oolite Group (Sellwood and McKerrow, 1974). This is typically a brown, fossiliferous non-oolitic, rubbly limestone with abundant shell debris and a marly matrix (Green, 1992). Again it was William Smith who coined the term 'Cornbrash', and within this rock type he first noticed that the fossils were distinctly different between the upper and lower layer. This led him to establish the principle of superimposition, a fundamental concept in geology, geomorphology and earth history.

Douglas and Arkell (1928) and Arkell (1933) divided the formation into upper and lower divisions based on brachiopods and to a lesser extent ammonites (Table 3). The lower division is of Bathonian Age and the upper belongs to the Callovian stage. The lower Cornbrash is thin, 6-7m, whilst the Upper Cornbrash is considerably thinner, with a maximum thickness of 3m (Childlaw and Campbell, 1988). The Cornbrash Limestone Formation is thickly bedded when fresh but, weathers to hard rubbly limestone at outcrop.

Table 3 Biostratigraphical divisions of the Cornbrash Formation (based on Arkell, 1933)

Ammonite zones		Brachiopod zones	Stratal division
Macrocephalites macrocephalus	*Rhynchonelloidea cerealis* group	*Microthyris lagenalis* *Microthyris siddingtonensis*	Upper Cornbrash
Clydoniceras discus	*Kallirhynchia yaxleyensis* group	*Ornithella obovata* *Cererithyris Intermedia*	Lower Cornbrash

Cretaceous and Tertiary deposits are absent across the Cotswolds. There are no remnants (apart from a few scattered flints) of the great Chalk cover that is thought to have mantled most of the British Isles, and no mantles of sands, clays and gravels that are such a feature of the Tertiary basins of southern England

2.4 Structure

In the Cotswolds, the regularity of the rocks has been disturbed by flexing and faulting (Horton *et al.*, 1987). In general these hills consist of faulted cuesta-blocks which dip gently towards the south-east and present steep-sided north-west facing scarps. The margins of these blocks are usually cambered and are separated by deeply incised valleys (Horton *et al.*, 1987). In some areas, the Lower Jurassic or Liassic formations are exposed. The main areas of scarp and higher ground consist of Middle Jurassic limestones, comprising Inferior Oolite, Fuller's Earth, Great Oolite, Forest Marble, and Cornbrash.

The main structures affecting the rocks of the Cotswolds are a number of small subsurface domes aligned in a roughly east-west direction, which are crossed by somewhat later east-north-easterly faults. These structures are additional to the gentle regional dip to the south-east, but their precise age is unknown. Horton *et al.* (1987) suggest that they may have been produced during either of the two widespread post-Jurassic periods of earth movement that have affected the rocks of southern England, i.e. early Cretaceous or mid-Tertiary. Extensive cambering has obscurred many faults, particularly in the Inferior Oolite and Marlstone Rock Bed. These add to the gentle regional dip in the south-west.

Hull (1857, p.75) recognised that there are two main fault systems, the commoner trending 12° north of west, the other north to south. They were probably formed during the Miocene orogeny (Benison and Wright, 1969; Hart, 1976).

Hancock (1969) discussed the structure of the Cotswolds but did not address the effect of joints and fault-swarms on the relief. The regional dip of the Middle Jurassic sediments is 1° to 2° to the east or south-east and this is crucial in determining the general form of the escarpment. Two principal fault directions are recognised, 10° north of west and north-south. The E-W trending faults have longer traces and are commoner. They are on the whole thought to be normal faults with an average throw of 20m in each conjugate system.

Three gentle fold-like structures occur in the Inferior Oolite

1. Painswick syncline
2. Birdlip anticline
3. Cleeve Hill syncline

There are no measurable differences of dip across them. In the field they show up by the variation of bed thickness and the effects of the unconformity as shown by the preservation of Middle Inferior Oolite in the two synclinal areas. As well as the above three fold structures there are two others which are Tertiary folds:

4. Winchcombe anticline
5. Blockley syncline

The location of these five structural features is shown in Figure 8. The angle and direction of the dip of the rocks and the development of the folds within them are, therefore, to be regarded as a legacy of the Tertiary orogeny. There are however, two curious features which require reconciliation in order to substantiate this view.

The first, a general problem not confined to the Cotswolds, was noted by Sparks (1962), which concerns the difficulties arising from the distribution and intensity of the uplift, which was greatest nearest the Alps. The second concerns the orientation of supposed mid-Tertiary folds in the Cotswold Hills (Hart, 1976). It is well known that most of the large mid-Tertiary folds in south-east England are aligned approximately E-W. This is, however, not the case in the Cotswolds. The Moreton and Winchcombe anticlines and the intervening Blockley syncline are aligned roughly N-S.

The Vale of Moreton anticline, orientated predominantly N-S, is regarded as a continuation of the Pennine line by Richardson (1925) and Cox and Trueman (1920). Rastall (1925) inferred that the Sedgley-Lickley axis, when continued through the Vale of Moreton, crosses this line. The significance of the axis lies in its position close to the margin of the troughs in which Mesozoic sediments were deposited. The axis was able to accentuate the effects of the defective deposition. Closely allied in direction, and possibly in origin too, is the NNW-SSE axis, supposedly an anticline, near Banbury.

Hart (1976, p.15) has addressed some of the effects of the structure in relation to relief. He suggests that

> "...the best example of the influence of faulting on relief occurs near Stow-on-the-Wold, however. Here the continuity of the Vales of Moreton and Bourton has been interrupted by a line of hills bounded to the north and south by E-W faults, forming a fault block."

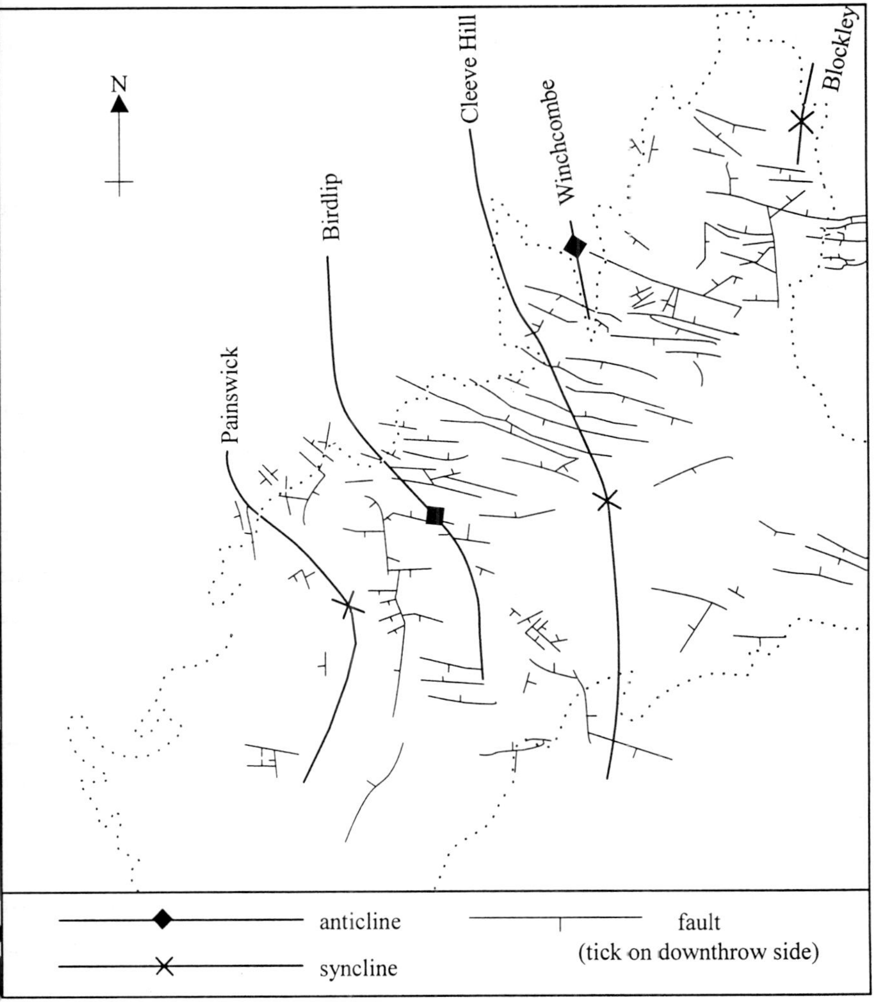

Figure 8 Structural features of the Cotswolds (after Hancock, 1969)

Hart also describes the effects of the anticlines and synclines and their relation to inverted relief, which were first noted by Hull (1855). For example, the Moreton and Winchcombe anticlines were denuded to form the Vale of Moreton and Winchcombe embayment and the intervening

Blockley syncline forms the main hill mass of the north Cotswolds. Hart (1976, p. 16) points out

> "It should be noted though the Vale of Moreton is more than a simple example of inverted relief developed on a Miocene anticline, because the line of the anticline corresponds to that of the Jurassic Moreton swell where disposition of the scarp forming Inferior Oolite was deficient, facilitating the denudation of the anticline."

Chapter 3

3.1 Quaternary deposits

The Quaternary Period, covering the last two million years, is divided into the Pleistocene and Holocene, the latter accounting for the last 10 000 years and equates approximately to the the period since the end of the last ice age (Goudie, 1983).

The Quaternary deposits of the Cotswold region comprise varied unconsolidated beds collectively known as 'Drift'. These include glacial, fluviatile and a mixed group of periglacial deposits known as 'Head', which grade into colluvium, mudflow and landslip (Green, 1992). The outstanding feature of the Quaternary has been the fluctuations in climate between cold, glacial and warm, temperate stages. Correlation of terrestrial, Quaternary deposits is made difficult by considerable lateral variations, hiatuses, lack of fossil evidence and exposures and problems with dating deposits. Despite these problems the Quaternary sediments of the Cotswolds offer perhaps, the longest terrestrial sequence in Britain (Bridgland, 1994; Parker, 1995). Interpretations of longer term progressive metamorphosis of the Cotswolds during the Quaternary has been aided by morphological, sedimentological, stratigraphic, palaeoecological and chronological information.

The Quaternary deposits of the Cotswolds have been studied for over one hundred and fifty years (Goudie and Hart, 1975). Dean Buckland (1823) discussed the curious quartzitic gravels of what he called 'The Northern Drift' in his catastrophic *Reliquaie Diluvianae*. He saw these materials as relics of Noah's Flood.

The Cotswolds and Upper Thames Valley lay outside the southern limit of the late Quaternary ice-sheets; thus, this area has escaped the direct effects of glaciation during at least the last two glacial episodes. Pleistocene deposits from earlier glaciations are, however, evident. These are heavily dissected and occur across the higher ground in the region. Similarly, this region lay beyond the reach of marine transgression during interglacial phases, and thus, it has not been affected directly by changes in sea-level (Briggs *et al.* 1985). Throughout the Quaternary, the Upper Thames basin has owed much of its development to fluvial action and associated hillslope processes. These processes have led to the formation of extensive river gravel terraces and channel infills (Macrae, 1985). For a considerable portion of the Quaternary the Cotswolds would have been subjected to the full rigours of a periglacial and tundra environment underlain by permafrost.

On the Cotswolds, widely scattered erratics, notably quartzite, known as remanié drifts exist. These are the earliest Quaternary deposits in the region. Often they have been included within the concept of 'Northern' or 'Plateau' Drift (Hull, 1855; Arkell, 1947a; Kellaway *et al.*, 1971) and they are believed to be pre-Cromerian in age (Horton *et al.*, 1987). Younger Quaternary deposits are confined to the valley sides and floors, where head, brickearths, peat, calcareous tufa, river-terrace deposits and alluvium are sporadically preserved.

The Quaternary deposits of the Oxford area were originally mapped by Pocock (1908) and subsequently evaluated by Sandford (1924, 1929, 1932, 1954 and 1965). Sandford mapped, interpreted and named the drift and terrace deposits of the Upper Thames, initially postulating the existence of four main terraces. He carried out his work at a time when originally four, and later five glacial episodes had been recognised, long before the ocean-coring programme (Shackleton and Opdyke, 1973 a,b) revealed a more complex record.

The Quaternary sediments of the Upper Thames basin were reviewed by Goudie and Hart (1975) and subsequently have received attention from Gilbertson (1975), Briggs (1975), Briggs and Gilbertson (1980), Briggs *et al.* (1984), Maddy *et al.* (1991), Whiteman (1991) and Bridgland (1994). The deposits of the Upper Thames itself are still poorly understood and considerable controversy surrounds the terrace sequence. Maddy *et al.* (1991) discuss the great confusion over the assignment of deposits to the various levels of the Thames and its tributary valleys. Few terrace deposits have been dated with any certainty and only the Floodplain Terrace is within the limits of radiocarbon dating. Therefore, there is uncertainty arising from difficulty in identifying, dating and correlating interglacial deposits in this area. Table 4 shows a likely lithostratigaphic classification of the Upper Thames Quaternary deposits. Figures 9, 10 and 11 show the terrace levels within the region.

3.2 Plateau or Northern Drift

The Plateau, or Northern Drift of the Upper Thames valley represents the highest and therefore the oldest Quaternary deposit in the region. It occurs over a dissected north-east to south-west belt extending from the Moreton to the Goring Gap at a height of 198-63 mOD (Coe, 1993).

This belt is approximately 13km wide and the deposits are generally less than 3m thick (Arkell, 1947a). They are less widely distributed than those of the younger Pleistocene (see Figures 9 and 12), being found mainly in the valley of the River Evenlode, a northern tributary of the River Thames and near the Thames itself between its confluence with the Evenlode and the Goring Gap. Outside this belt they tend only to be present as scattered erratics or as crevice fillings.

These deposits are generally considered to consist of de-calcified, non-local, quartz and quartzite-rich, rubefied red sandy to clay-rich gravels which have most frequently been interpreted as marine, fluvial or glacial sediments over the last 150 years (Sandford, 1924; Arkell, 1947a; Goudie and Hart, 1975; Shotton *et al.*, 1980). Indeed, the origins of the Plateau Drift have for long been debated.

In the early nineteenth, scientists saw them as being relicts of the Noachian Deluge (e.g. Buckland, 1823), but by the later part of the nineteenth century they were being interpreted as either of fluvial or of glacial origin. Davis (1899), following White (1897) believed that they might originally have been carried by rivers connected with the Severn system across into the present catchment of the Thames before the Thames catchment was beheaded by the retreat of the escarpment and the diversion of the former eastward flowing streams down the developing Severn Vale. On the other hand there is a considerable height range (135m), with facies variation, lack of stratification and lack of local lithologies, and there have been cited as evidence of a glacial origin. The bulk of the material (*ca.* 85%) is made up of material of Bunter origin from the English Midlands.

The distribution, origin and topographic disposition of the Plateau Drift have recently been reviewed by Shotton *et al.* (1980). The age and origin of these materials are subjects of considerable debate. The oldest deposits which can be correlated fill a channel at Sugworth near Abingdon and consist of organic deposits which pass laterally into gravels of Plateau Drift type. Their rich flora has been assigned to the late Cromerian. The channel is covered by a veneer of gravel of Plateau-Drift type, thought to be a solifluction deposit. On this basis the Plateau Drift must at least be in part pre-Cromerian in age or at least older than half a million years (see Table 4).

Group	Sub-Groups	^{18}O	Stage
Recent floodplain and channel deposits(alluvium, peats, tufas)		1	Holocene
Floodplain Terrace	1a. Northmoor	2-4	Late Devensian
	1b. Sutton Courtenay	5a-5c	Early/mid Devensian
Summertown-Radley Terrace	Eynsham Gravel	5e	Ipswichian
	Stanton Harcourt Gravel	6	Late-Saalian
	Stanton Harcourt Channel	7	Intra-Saalian temperate stage
	Basal Summertown-Radley Formation	8	Mid-Saalian
Wolvercote Terrace	Wolvercote Gravel	8	Mid-Saalian
	Wolvercote Channel	9	Intra-Saalian temperate stage
	Basal Wolvercote Gravel	10	Early Saalian
Hanborough Terrace	Moreton Drift?	10or12	Anglian/ Hoxnian
	Hanborough Gravel	10-12	Hoxnian
Plateau Drift	Freeland Gravel	12	Anglian
	Sugworth Channel	13-21	Cromerian
	Coombe Gravel	pre-21	Early Pleistocene
	Wilcote Gravel	pre-21	Early Pleistocene
	Ramsden Heath Gravel	pre-21	Early Pleistocene
	Waterman's Lodge Gravel	pre-21	Early Pleistocene
	Nettlebed Organic	pre-21	Early Pleistocene

Table 4 Lithostratigraphic classification of the Upper Thames Quaternary deposits (after Bridgland, 1994)

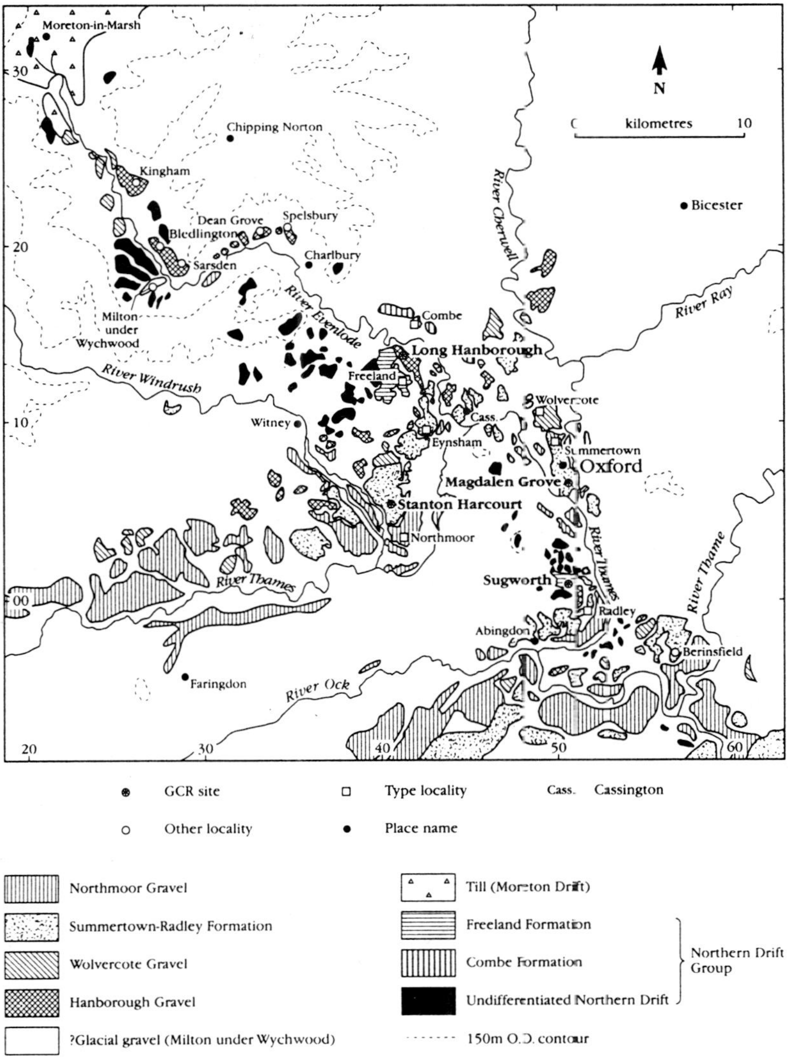

Figure 9 The gravels of the Upper Thames catchment (after Goudie, 1976; Bridgland, 1994)

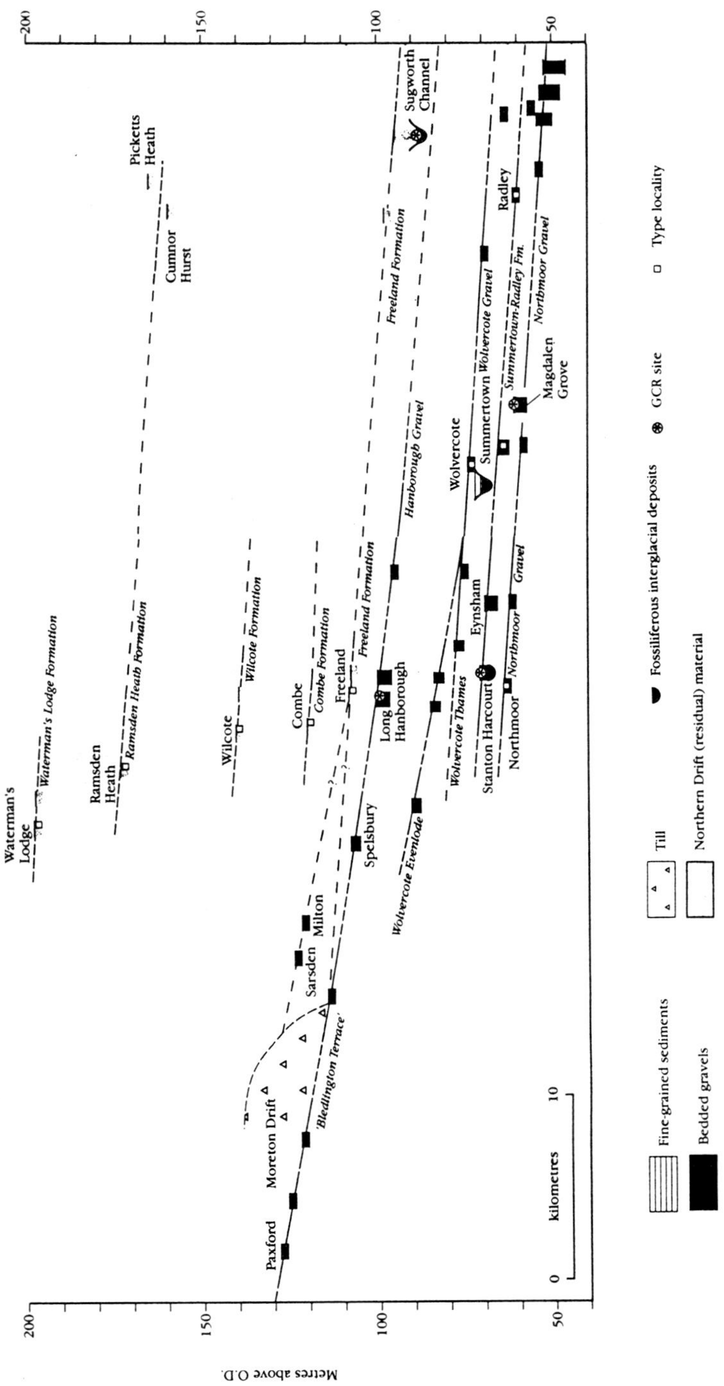

Figure 10 Longitudinal profiles of the Upper Thames terrace surface (after Bridgland, 1994)

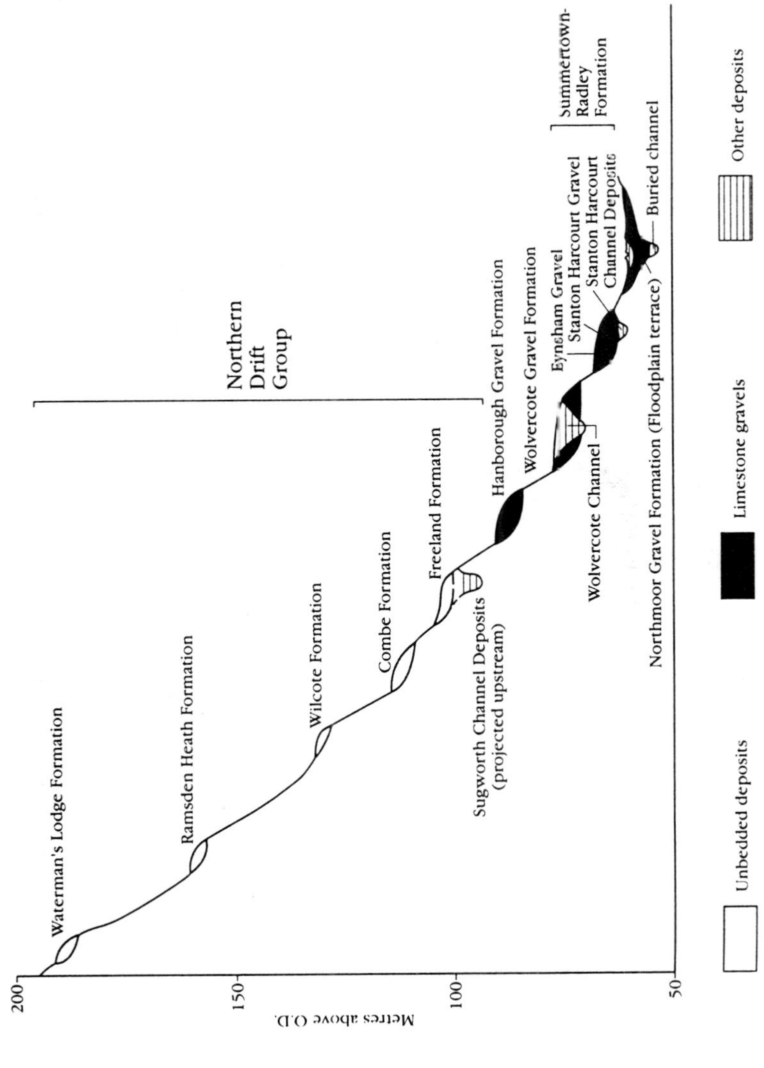

Figure 11 Idealised transverse sections through the terrace deposits of the Upper Thames (after Bridgland, 1994)

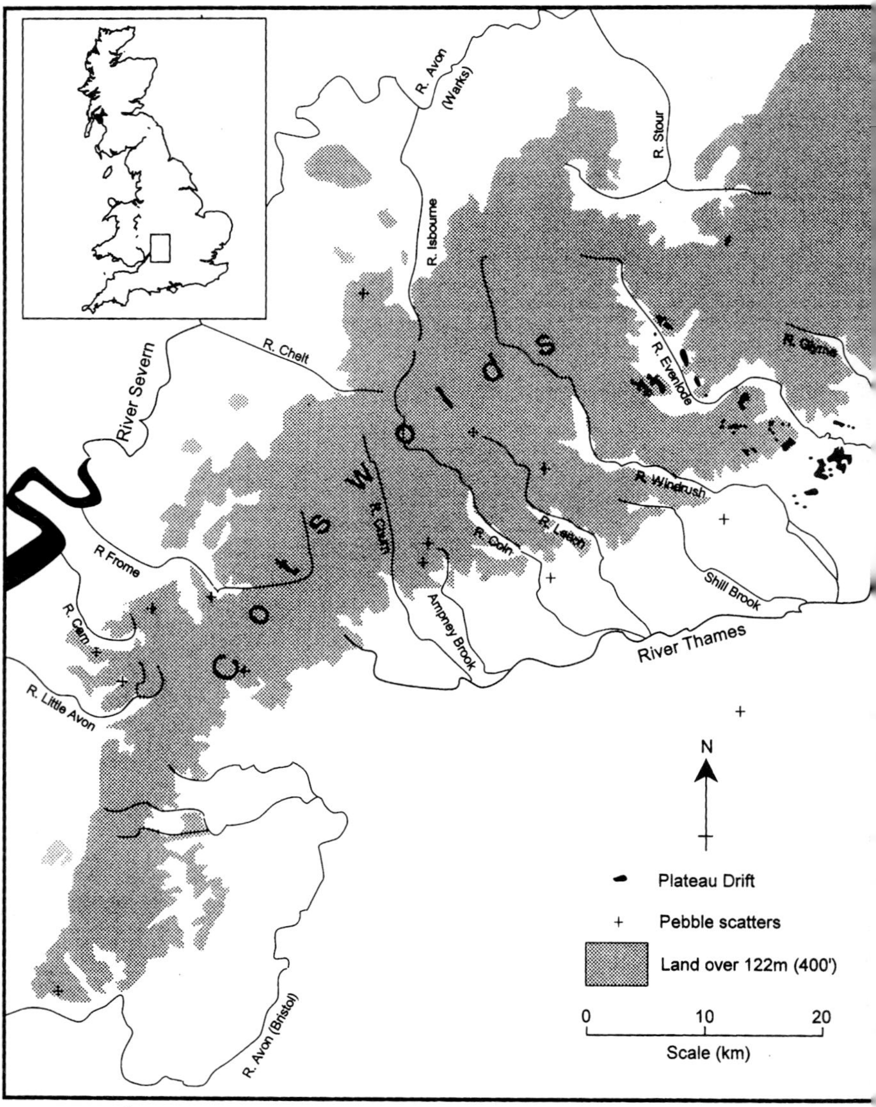

Figure 12 The distribution of Plateau Drift in the Cotswolds

Since the discovery of the Sugworth site, Hey (1986) proposed an alternative interpretation of the Plateau Drift. Hey divided the Plateau Drift into individual gravel aggradations based on elevation and composition. These are as follows: Waterman's Lodge, Ramsden Heath, Wilcote, Coombe and Freeland Members (as shown in Figures 10 and 11). Hey disputed the glacial origin of the materials and suggested that they are the remains of old river terrace gravels, using surface pebble counts as his main line of argument. His ideas are supported by a number of workers including Bowen *et al.* (1986), Bridgland (1988) and Whiteman (1991). Goudie and Hart (1975) on the other hand suggested that the most likely mechanism is a glacial origin on the basis that the deposits are an ill-sorted mixture of clay and pebbles alien to the Cotswold bedrock lithology, The material lacks stratification, many of the pebbles are broken, requiring at least 25 KN pressure to fracture the quartzite Bunter pebbles (Coe, pers comm.) and have been reported as showing parallel striations which are caused during glacial entrainment. Goudie and Hart (1975) suggest that the marked and rapid changes of facies are best explained by glacial and fluvio-glacial energy. Recent work carried out by Bull and Goudie (1987), including scanning electron microscopy, and Coe (1993 and pers comm.) suggest that the Plateau Drift is not a simple or single deposit, but includes materials of fluvial and glacial origin, as well as severely reworked soliflucted materials.

In the southern Cotswolds, Ramsay *et al.* (1885, p45) discussed the occurrence of scattered erratic pebbles from the Cotswolds between Nailsworth and Bath. These high level plateau deposits comprise poorly sorted, loamy gravels with abundant Cretaceous flints and cherts along with Jurassic and Upper Palaeozoic material and are considered as 'Plateau or Northern Drift' (Green, 1992). Sites at Uley Bury (SO 785 989), Wotton-under-Edge (SO 789 963), Nympsfield (SO 810 008), Minchinhampton Common and Tetbury were recorded (Lucy, 1872, p.108, 1880, p.51-52). On Bathampton Down (over 200 mOD) Plateau Drift material was first described by Varney (1921) and later by Richardson (1955). Hawkins and Kellaway (1971) found Plateau Drift material in fissures in the Great Oolite. Although of a similar nature, the relationship between the main deposits of the Northern/Plateau Drift in the southern with that from the northern Cotswolds is not yet clear. However, the scattered nature of the deposits in both areas indicates both their antiquity and the substantial degree of post-depositional modification that has occurred.

3.3 Vale of Moreton deposits

The Anglian glaciation, which perhaps dates back to *ca.* 400,000 years ago, is better represented in the Northern Cotswolds in the Vale of Moreton (Sumbler, 1995). There would appear to be a three-fold sequence. At the base lie the Stretton Sands (incorrectly ascribed as being the Cheltenham Sands in this area by Richardson (1937a). Though of a similar nature, they are much older in age). These are fluviatile, cross-bedded quartz sands, which have yielded vertebrate remains of temperate affinity including straight-tusked elephant and red deer which have probably been derived from an earlier interglacial (Williams and Whittaker, 1974). Overlying the Stretton Sands are the Paxford Gravels, which comprise local Jurassic limestone material. Richardson and Sandford (1963a) record the remains of mammoth from the Paxford Gravels from Ditchford. They note that the contact between the Paxford Gravel and underlying Stretton Sand has an irregular surface and appears to have been erosive. On top of the Paxford Gravels lie several metres of Chalky Boulder Clay, generally referred to as the Moreton Drift. This comprises locally abundant fresh flints and glacial sands and gravel with derived Bunter pebbles (Richardson, 1937a; Richardson and Sandford, 1963a). At the contact between the Paxford Gravel and the Chalky Boulder Clay a thin red-clay (Wolston Clay) is locally present. This has been postulated as being a feather-edge remnant of the glacial deposits of glacial Lake Harrison, a possible glacially dammed lake (Bishop, 1958; Williams and Whittaker, 1974). Several cols in the Jurassic escarpment have been interpreted as overflow routes from this lake (Shotton, 1953; Bishop, 1958). Dury (1951) mapped the supposed shoreline of the lake on the Jurassic scarp in the form of a narrow bench at 125m (400 ft). Ambrose and Brewster (1982) suggest that the feature is in fact caused by an outcrop of a slightly more resistant bed in the Lower Lias, and not a wave cut notch.

Between Ebrington Hill and Dovers Hill orange quartzose sands similar in nature to, but thought to be much younger than the Stretton Sands, may be found. These are known as the Campden Tunnel Drift. The deposits occupy what is believed to be a glacial overflow channel (Bishop, 1958), caused by the ponding of the Avon and Severn Valleys by the Welsh glacier farther down stream (see Figure 13).

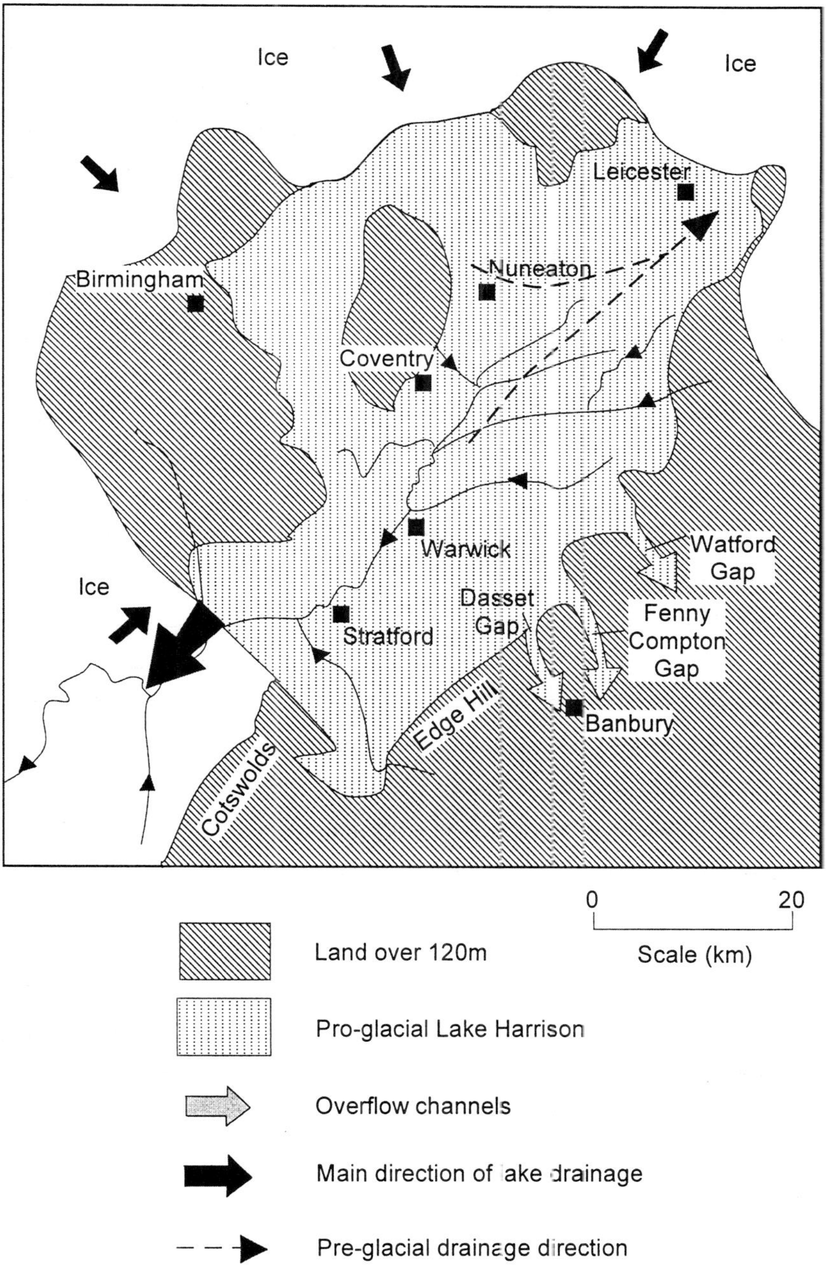

Figure 13 The extent of the pro-glacial Lake Harrison in the Midlands (after Shotton, 1953)

3.4 The overall impact of glaciation on the Cotswolds

Although the presence of spreads of gravel in the Cotswold region may be indicative of an early past glaciation, and although ice certainly entered the Vale of Moreton, the direct impact of glaciation in the Cotswolds is very modest. Indeed, there is no evidence for glacial erosion or deposition over most of the region. There is, for example, no wholesale modification of the scarp form of the type that was achieved in the northern part of the Cretaceous escarpment between Dunstable and Hunstanton, nor is there a blanket of Boulder Clay smothering the landscape. On the other hand there is a possibility that glacial action had some significance. Kellaway *et al.* (1971), for instance, suggested that ice sheets overran most of the Cotswold Hills, that the upper part of the Thames Valley must have been buried beneath thick ice, and that the broad gap between Bredon Hill and the Cotswolds might have been excavated in whole or in part by ice.

3.5 Terraces of the River Thames

3.5.1 Hanborough Terrace

With the exception of the Coombe and Freeland Terraces (Arkell, 1947a) the Hanborough Terrace is the highest and oldest of the limestone dominated river-gravel accumulations recognised by Sandford (1924, 1929, 1932, 1954, 1965) and occurs at about 80 mOD in the Oxford area. The most extensive and well preserved spreads of the Hanborough Terrace occur in and around Long Hanborough (Figure 9). The Hanborough Terrace gravel differs considerably from the Plateau Drift in that it contains around 90% local materials with small quantities of quartz, quartzite, sandstone, flint and igneous and metamorphic rocks (Briggs and Gilbertson, 1973, Bridgland, 1994). Hanborough Terrace gravel spreads occur in the Evenlode Valley with notable deposits at Long Hanborough, Bledington, Cassington and Lyneham. In the Vale of Moreton they are to be found at Ditchford, Paxford and Stretton-on-Fosse, in the Windrush Valley at South Leigh and Hardwick, and in the Cherwell valley at Bletchington and Kirtlington. The terrace has a low gradient, and in the upper reaches of the Evenlode valley it is overlain and obscured by alluvium. The Hanborough Terrace gravels have yielded rolled vertebrate bones of elephant, auroch and red deer, which are normally associated with temperate forest conditions. Therefore, this led to the assumption that these were an interglacial accumulation, possibly related to high sea-levels (Sandford, 1924, 1926; Arkell, 1947a). However, cold stage molluscan faunas from channel deposits within the gravels led Briggs and Gilbertson (1980) to propose an early Wolstonian age for the unit. Furthermore, the sedimentary structures of the gravels and sands, and their association with periglacial structures indicate deposition in a cold climate, braided river environment. These gravels contain a small proportion of erratics presumably derived from the Plateau Drift. They are noted for their clay-filled pipes, which are thought to result from a combination of cryoturbation and solution. This terrace is believed to have accumulated under semi-arctic conditions on the basis of a cold-tolerant molluscan fauna with abundant *Pupilla muscorum*, possibly during the early part of the Anglian Stage with temperature fluctuations between -15 to +8°C suggested (Briggs *et al.*, 1985). The existence of a cold environment has been supported by a range of periglacial features including boils and cracks, carrot-shaped ice-wedge casts and involutions (Briggs, 1973; Briggs and Gilbertson, 1973).

3.5.2 Wolvercote Terrace

The Chalky Till ice-sheet, of debatable age, reached the headwaters of the Evenlode and Cherwell rivers but did not penetrate the present study area (Bishop, 1958). The proximal outwash may be represented by limestone dominated, flint-bearing gravels which comprise the Wolvercote Terrace of the Evenlode and Cherwell (see Figures 9, 10 and 11). Early investigations (Bell, 1904; Sandford, 1924; Tomlinson, 1929, Bishop, 1958), based on biological and sedimentological evidence, indicated that aggradation occurred under cold-stage, even glacial, conditions rather than under interglacial conditions and that the terrace was a product of outwash from this ice-sheet. Arkell (1947a) postulated deposition occurred at a time post-dating the Hanborough Terrace. The proportion of flint in the gravels decreases downstream, concomitant with an increase in the amount of limestone. An organic deposit occurs at Wolvercote (see below), but its exact relationship with the Wolvercote Terrace is uncertain (Sandford, 1924).

The terrace lies at an altitude of 67-70 mOD and is characterised by a greater frequency of flint compared with other limestone-rich gravels. A number of authors have suggested that this represents the discharge of material from the Wolstonian (Saalian) stage Chalky Till glaciation of the Midlands (Bishop, 1958; Tomlinson, 1963; Shotton *et al.*, 1973; Briggs and Gilbertson, 1974). Bowen *et al.* (1986) and Rose (1987) have, on the other hand, suggested that it might be Anglian in age.

Recent work by Maddy *et al.* (1991) has challenged the existence of the Wolvercote Terrace. Their work has suggested that the Hanborough Terrace and Wolvercote Terrace are on lithological grounds the same, though this still has to be tested.

At the turn of the century, a large channel filled with a sequence of gravels, sands, peats and silty clay was revealed at Wolvercote (SP 498 105) in a brick pit, cut into the underlying Oxford Clay. These sediments yielded faunal and floral remains which indicated a cooling phase during fully temperate conditions (Bell, 1894, 1904; Sandford, 1924, 1926; Arkell, 1947a, Duigan, 1956), as well as a large number of Palaeolithic artefacts (Wymer, 1968; Roe, 1981; Tydelsley, 1986a,b, 1988).

There have been contrasting interpretations with both Hoxnian and Ipswician ages proposed (Bishop, 1958; Tomlinson, 1963; Shotton *et al.*, 1973). Sumbler (1995) has sought to correlate it with the Blackditch Terrace of the Thame and the Lynch Hill Terrace of the Middle Thames sequence.

3.5.3 Summertown-Radley Terrace

The Summertown-Radley Terrace is the most extensively preserved terrace of the Upper Thames (Figure 9). It stands 6-10m above the Floodplain Terrace, and consists largely of limestone with minor quartzite and flint gravels.

Sandford (1924, 1926, 1954) suggested two components; a lower part with a cold fauna, which includes mammoth, woolly rhino and bison, and an upper part with a warm fauna including hippo and elephant. Arkell (1945) suggested a more complex sequence, interpreting early human (Acheulian) implements collected from the lower part of the gravels as indicative of interglacial conditions. He thus related this to a tripartite warm-cold-warm sequence.

Recent work has suggested as many as five distinct units within the terrace, a complex sequence spanning two temperate episodes (including a buried channel, described below) and a full cold interval with parts of two others (Bridgland, 1994). The cold part, named the Stanton Harcourt Gravel, is by far the most widespread. The interglacial deposits appear to be confined to relatively localised channels and to zones of reworked materials within the terrace gravels. Channel deposits have been suggested as pre-Ipswichian in age on the identification of preserved fossils (Bridgland *et al.*, 1989). This poses a major problem which has not been widely accepted (Gibbard, 1985). It pre-supposes an as yet unrecognised post-Hoxnian/pre-Ipswichian interglacial.

A shallow, broad channel at the base of the Summertown-Radley Terrace was revealed during gravel extraction at Linch Hill near Stanton Harcourt. The channel was incised into Oxford Clay, and filled with a series of gravels, sands, silts and peats. A varied vertebrate fauna was found and a species-rich molluscan fauna was present (Briggs *et al.*, 1985). The fauna and flora from this channel, along with Acheulian ovate handaxes, suggest fully interglacial conditions (Bridgland, 1994). Seddon and Holyoak (1985) attempted to date the channel material with conflicting results. The basal material at Stanton Harcourt was dated to 34 700 years BP (implying a mid-Devensian age). Thermoluminescence dates yielded ages between 91 000 and 93 000 yrs BP derived from silts higher up the same channel sequence.

The terrace is composed of medium-to coarse-grained, sub-rounded to sub-angular limestone gravel. Like the Hanborough Terrace, this cold-stage aggradation may be of fluvio-periglacial origin (Goudie and Hart, 1975).

The upper Summertown-Radley sediments have been named the Eynsham Gravel (Briggs *et al.*, 1985). These represent temperate deposits (Sandford, 1924; Dines, 1940) confined to a distinct phase.

Continuing excavation of the Summertown-Radley Terrace across the region will increase the data available from these terrace deposits.

3.5.4 Floodplain Terrace

The youngest Pleistocene river deposits of the Upper Thames form the Floodplain Terrace. Its gravels create extensive spreads, many of which extend beneath the recent alluvium. The Floodplain Terrace is composed almost entirely of local materials. Jurassic oolitic limestone comprises about 90% of the gravel, while ironstones, mudstones, and marlstones from surrounding local sources make up most of the remainder. Small quantities of rock from outside the region also occur, including flint from both the Berkshire Downs and the Chalky Till of the Vale of Moreton (Bishop, 1958). In addition, minor proportions of quartz, quartzite, igneous and metamorphic rocks are found in the terrace sediments, presumably derived from the older Plateau Drift (Sandford, 1924).

The British Geological Survey suggest that two distinct surface levels occur on the Floodplain Terrace. This terrace also has two members, with a vertical separation of 0.3-0.5m. These are designated terrace 1a (Northmoor) and 1b (Sutton-Courtenay), 1a being the lowest and thus, closest to the present river (Robson, 1976; Harries, 1977). Although the terrace has two distinct members, it is not clear whether these represent two climatic phases. In the upper reaches of the Thames and tributary valleys, the terrace surface also varies in elevation laterally. The gravels rise onto the flanks of the valley in places, grading into older deposits. As a result difficulties are encountered in correlating the various terrace flats. Variations in the terrace morpohology and sedimentology between the higher and lower reaches of the Upper Thames indicate that fluvial conditions were not uniform throughout the catchment.

The fauna, flora and cryoturbation structures within the deposits indicate accumulation under very cold climatic conditions, with gravel aggradation in a broadly cold environment, as shown by intra-formational ice-wedge casts and cryoturbation layers in the terrace gravels. The main constituents are medium to coarse limestone gravels with an interstitial fraction of medium-grained sand, and the sediments are similar to those found in braided streams. It is almost impossible to assess the quantitive hydrological implications of these terrace deposits, but Briggs (1983), using a variety of palaeohydrological approaches, has made general order of magnitude estimates of discharge, indicating peak flows at least 13 times greater than the present discharge.

Variations occur in the bedrock surface beneath the Floodplain Terrace gravels. These variations led Sandford (1924) to postulate the existence of a buried channel beneath the terrace gravels. Castleden (1977) has argued subsequently that the supposed channel is no more than the centre of a wide, periglacially-trimmed valley floor, but Robson (1976) and Harries (1977) showed that abrupt variations in the sub-drift land surface do occur, implying the presence of a relatively complex bedrock topography.

Of the main terraces, the Floodplain Terrace is the only one which can be accurately radiocarbon dated. The Northmoor Terrace of Arkell (1947a) has been shown by Briggs and Gilbertson

(1975, 1980) to include both middle and late Devensian aggradations. Radiocarbon dates from organic channel fills fall into two groups: 40 000 to 29 000 yr BP and 13 500 to 10 500 yr BP (Briggs and Gilbertson, 1980; Aalto *et al.*, 1984). Briggs (1973) associated the Sutton Courtenay with the Upton Warren interstadial (mid Devensian age) and the Northmoor with the Allerød (late Devensian interstadial age). A brief review of the sites which record these events will now follow. It should be noted that some of these sites lie just on the margins or outside the limits of the Cotswolds (*sensu stricto*), but are included so as to provide a general picture of environmental conditions and changes during the Devensian.

At Poole's pit, Lechlade (SP 215 013), the lower deposits of the terrace gravels were laid down *en-masse* either by solifluction, or by sediment choked streams which were able to transport the excessive quantities of debris entering the system (Briggs *et al.*, 1985). The upper gravels on the other hand, indicate a more active fluvial system, increased entrainment and transport, and improved sediment sorting. Palaeoecological analyses from a peat lens at the base of the terrace were undertaken by Briggs *et al.* (1985). The peat yielded a radiocarbon date of 28 400±1000 yr BP (Birm-1170). Pollen from this sample contained a range of herbaceous taxa, indicative of cold conditions, with high values of Compositae, *Plantago*, *Ranunculus* and Gramineae present. Low values of *Filipendula*, *Betula* and *Juniperus* were also noted. Coleopteran and Molluscan remains indicated a similar picture of open, cold conditions.

In the Vale of Bourton, large tracts of Floodplain Terrace gravels have accumulated, especially between the confluences of the Rivers Windrush, Dikler and Eye. Gravel extraction over the past 75 years has yielded numerous vertebrate remains including *Elephas primigenius*, *Trichorinus antiquitatus* but especially *Mammathus primigenius* (Sandford and Richardson, 1961; O'Neill and Shotton, 1974). At Little Rissington, an elephant tusk 3.3m below the present surface was dated to 34 400±800 yr BP (Birm-466) (O'Neill and Shotton, 1974). It should be noted that Briggs *et al.* (1985, p.32) incorrectly cite this date as 34 500±800 yr BP. Richardson and Sandford (1960) discovered remains of *Mammatus primigenus* and *Coelodonla antiquitatus* near Bourton-on-the-Water. Briggs *et al.* (1985) identified non-marine molluscan faunas from Bourton, which contained a mix of xerophile and hygrophile species characteristic of mid Devensian floodplain environments, which lacked warmth indicators.

At Thrupp Farm House Quarry near Abingdon (SU 452 059), a Late-Devensian organic fill in the sand and gravel of the Floodplain Terrace was exposed during commercial extraction. Aalto *et al.* (1984) examined the deposit, which probably accumulated in a bed-load dominated braided stream, using sedimentological and palaeoecological techniques. Two radiocarbon dates of 13 260±120 yr BP (Hel-1092) and 13 580±120 yr BP (Q-2017) were obtained from an organic clay, and indicate accumulation of this deposit during the Late-Devensian substage, which was characterised by a predominatly cool or cold climatic regime. From the work undertaken, it was suggested that the climate deteriorated during the deposition and accumulation of this deposit.

At Brown's pit, Linch Hill, a series of organic layers from the full and late Devensian periods was revealed during sand and gravel extraction from the Northmoor (Floodplain) Terrace. The flora and molluscan fauna have previously been described (Sandford, 1965). Pollen was obtained from a coarse detritus mud which contained many monocotyledonous plant fragments. The analysis showed a high frequency of non arboreal pollen, suggesting open, relatively tree-less conditions at the time of deposition. *Betula* was the main arboreal pollen type represented, with a low count of *Pinus* as well. A high frequency of *Salix* pollen was recorded (25%), suggesting marginal carr vegetation, along with the *Betula.* Marginal marsh species were represented by pollen of Cyperaceae, *Filipendula* (meadowsweet), *Thalactrium* (rue), *Menyanthes* (bogbean) and *Sparganium* (bulrush). Macroscopic plant remains also suggesting carr or marginal marsh vegetation surrounding the water body were recorded, including *Filipendula ulmaria, Mentha aquatica* (water mint), *Thalactrium flavum* (rue) and *Urtica dioica* (nettle). Pollen of Gramineae and macro-fossils of *Cerastium* cf. *arvense* and *Linum anglicum* suggest dry conditions and open grassland. Peat from these silt lenses has been radiocarbon dated at 11 250±100 yr BP (Birm-105) by Shotton *et al.* (1970), suggesting they were deposited during the Late Glacial Interstadial.

Excavations at Farmoor (Coope, 1976; Lambrick and Robinson, 1979) yielded environmental information from various periods. The earliest deposit analysed was a late-Devensian palaeochannel cut-off on the Thames floodplain. Fruits of *Betula nana* (dwarf birch) and calyces of *Armeria maritima* (thrift) were found; these are indicative of a park tundra environment. The peat deposits also contained a cold climate Coleopteran assemblage and were radiocarbon dated to 10 600±250 yr BP (Birm-590) (Coope, 1976).

At Mingies Ditch (SP 391 059) in the valley of the R. Windrush, gravel extraction exposed numerous fluviatile organic layers and silt lenses (Allen and Robinson, 1993). Various layers and horizons and layers from this site have been examined for pollen and a 'composite' diagram has been constructed and is shown in Figure 14.

The earliest deposit examined (horizon G) was an organic silt layer from the bottom of a gravel-filled channel crossing the Devensian gravels of the lower Windrush/Thames floodplain. Pollen content was sparse but included *Betula, Pinus, Salix,* and *Juniperus* (AP and Shrubs) and Gramineae and Compositae-Liguliflorae. Twigs of *Betula* and *Salix/Populus* sp. from the sample gave a radiocarbon date of 10 860±130 yr BP (Har-8356).

Excavations for a new sewerage pipeline between Great Rissington and Bourton revealed a series of organic lenses overlying fluvial floodplain terrace gravels near Great Rissington. An organic silty lens in the gravels yielded wood remains indentified as *Cornus*, which was radiocarbon dated to 10 710±100 yr BP (Wilkinson *et al.*, 1994) Pollen from this silty lens revealed an open flora characteristic of the late Devensian in southern Britain during this period. The local environment in the vicinity of the river was a *Filipendula* dominated fen-carr. Evidence for eutrophic pools were given by the presence of *Myriophyllum, Equisetum, Typha* and *Sparganium*. Disturbed soils were suggested by the presence of *Rumex* and *Artemisia*. A sparse tree cover was evident with small quantities of *Salix* and *Betula* present.

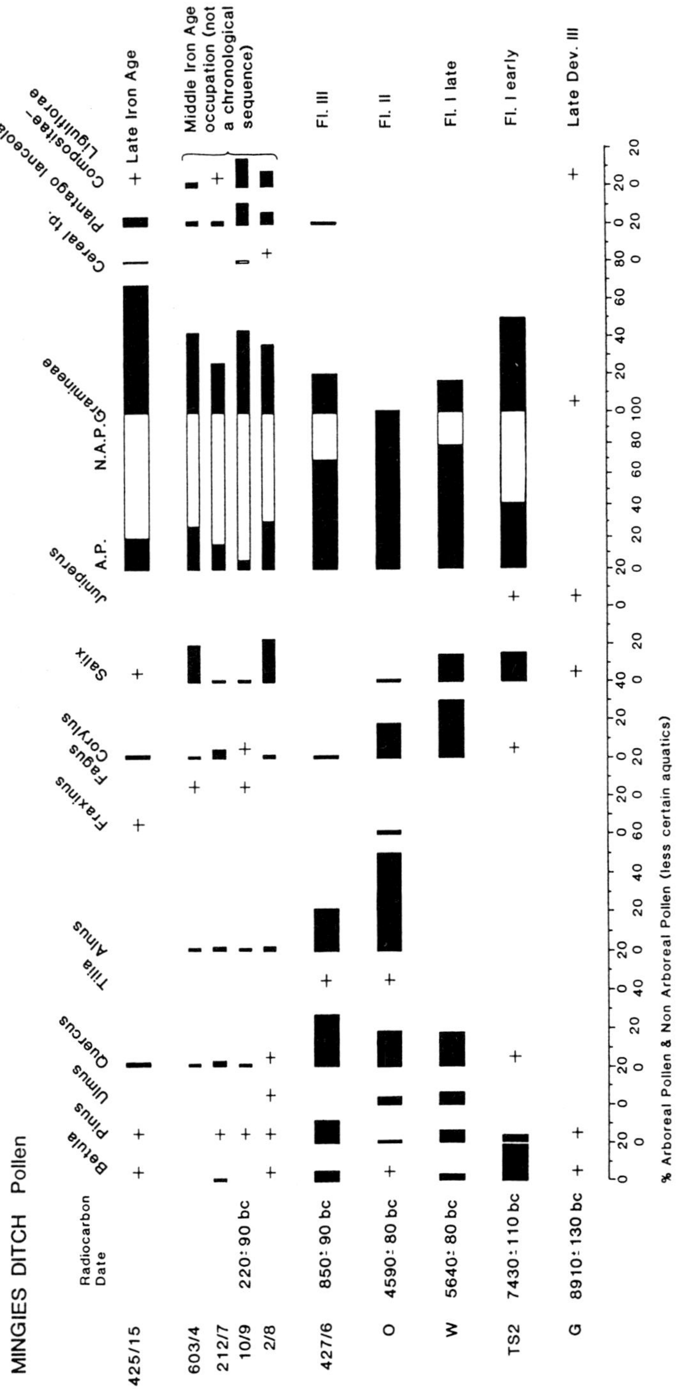

Figure 14 Composite Pollen Diagram from Mingies Ditch (after Allen and Robinson, 1993)

3.6 Terraces of the River Frome

The valley of the River Frome (or Stroud Water) begins *ca.* 100m south of the Witcombe embayment of the Cotswold escarpment. The valley runs southwards to Sapperton, where it sharply turns eastwards and deepens. The valley breaches the Cotswold escarpment a few kilometres south of Gloucester, opening out onto the Severn Vale (Large, 1963; Murray and Hawkins, 1973). This provides the outlet for the Painswick, Slad, Golden and Nailsworth Valleys which converge on its eastern end (see Figure 15a). These along with the Frome have unusually steep valley sides, giving this area its distinctive morphology. The River Frome has three distinct terraces (Large, 1963): the Whitminster, Cainscross and Stroud which shall now be discussed.

3.6.1 Whitminster Terrace

This forms the highest member of the series. At Stanley Down its altitude is 53 mOD, Nupend 40 mOD and Whitminster 33.5 mOD (Gardiner, 1938). Today, there are, few, if any satisfactory terrace exposures, and thus the evidence presented is that from former accounts. According to Gardiner (1931; 1938) and Large (1963) the gravel comprises medium-sized oolitic gravels intermingled with sand and silt. Within the terrace molluscs including *Lymnaea truncatula* and *Succinea pfefferi* have been recorded. Gardiner associated the Whitminster terrace with the No.4 Avon (Warwickshire) Terrace of Tomlinson (1929, 1941) and with the Kidderminster Terrace of the Severn of Wills (1938).

3.6.2 Cainscross Terrace

This is the best developed of the Frome terraces. Tomlinson (1941) noted that the Cainscross Terrace forms a curve steeper than that of the present river profile. Large and Sparks (1961) suggested that this was the result of a former high load/discharge ratio, under cold conditions.

The gravels consist almost entirely of rolled Jurassic Oolite and Lias material of local origin, with no recorded erratic material, along with coarse sands and silts. There is no uniform pebble orientation and much of the material is sub-angular to completely rolled (Large, 1963). A well developed bench-like feature occurs between Stroud and Ebley and is clearly shown in Figure 15b after Gardiner (1938).

A number of non-marine mollusca from within the terrace have been identified (Clifford, 1936; Gardiner, 1931, 1936, 1938; Large and Sparks, 1961; Large, 1963).

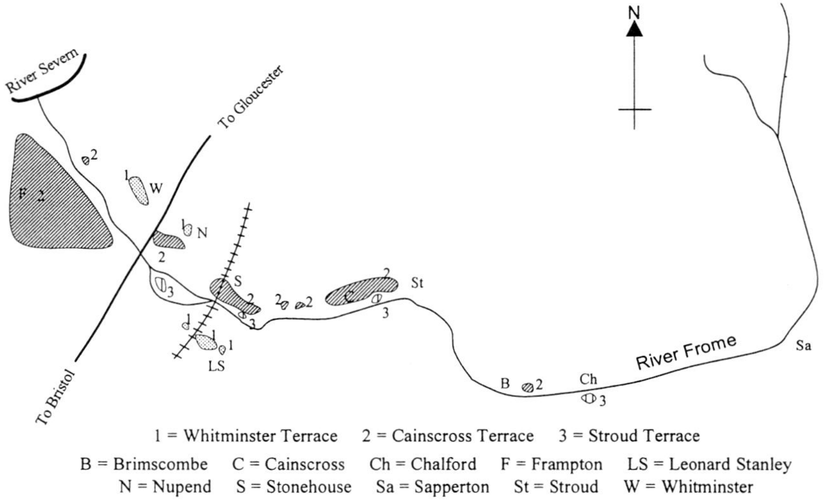

Figure 15a Map of the River Frome (after Gardiner, 1938)

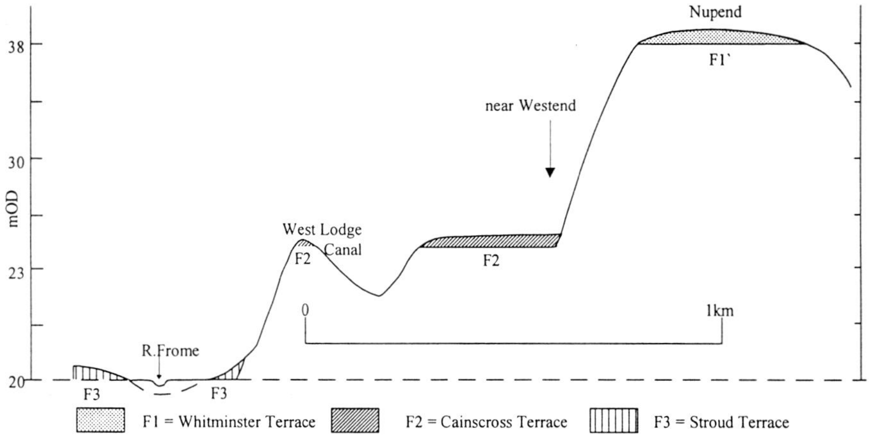

Figure 15b Section across the Frome showing the three river terraces (after Gardiner, 1938)

The molluscan faunas tend to be dominated by *Pupilla muscorum* along with smaller numbers of *Lymnaea truncatula*, *Succinea arenaria* and *Columella columella*. Large and Sparks (1961) suggest that the fauna on the whole was characteristic of cold conditions with possible loessic deposition, and were most probably of a Devensian age. Gardiner (1931) also noted the occurrence of a number of 'Dreikanter' stones, polished by sand which he attributed to aeolian conditions.

Sand lenses within this terrace consist of polished quartz grains, which Gardiner (1931) suggested as being derived from the Cheltenham Sands (see below). Large (1963) has also stated that this material is aeolian in origin, though not sufficient in quantity to justify a special explanation for its origin.

Within a number of exposures of the Cainscross Terrace, Tomlinson (1941), noted a channel infill of buff-coloured loams. She stated that

> "...the material in this channel is different from the deposits in which it is cut and appears to consist of an unbedded loam. This may represent a latter channelling of the Cainscross Terrace."

Large (1963), however, points out that the physical nature of this loam fill is aeolian in origin. A large number of vertebrate remains have, too, been found in the Cainscross Terrace, mainly from the Cainscross and Gannicox Pits (Gardiner, 1931, 1938). Finds include *Elephas primigenius*, *Rhinoceros trichornus*, *Bos* sp. and *Equus* sp. At Eastington Pit, a late Palaeolithic flint blade was found *in situ* within the gravels 7m above the present river (Gardiner, 1931).

3.6.3 Stroud Terrace

This terrace occurs beneath the present valley floor and is largely covered by alluvium. It is only very occasionally exposed. Gardiner (1938) described an exposure from the site of the Stroud Gas Works in which Stroud Terrace gravel was present 3.5m below the ground surface, resting on top of Lias Clay. Peat within the floodplain sediments of both the Stroud and Nailsworth valleys contained the remains of birch, ash, beech, pine, alder, hazel and horsetail (Playne, 1872; Gardiner, 1931). Vertebrate remains from the peat beds have included horse, ox, red deer, pig, boar and beaver (Playne, 1872; Gardiner, 1938).

Wills (1938) and Tomlinson (1941) have suggested that the Frome terraces are not normal fluvial terraces but owe their existence to periglacial activity. The work of Large (1963) has supported their ideas and he has suggested that the terraces are in fact solifluction deposits and not fluvial phenomena. Thus caution must therefore be adopted when trying to fit any chronology to them.

3.7 Terraces: general considerations

The terraces that mantle extensive areas of the lowlands that fringe the Cotswolds are for the most part, though not entirely, the consequences of cold climate conditions. It is likely that most of them were laid down by sediment-laden multi-channel streams. It is also probable that they derived their large sediment load as a result of intense frost-shattering, solifluction and high seasonal runoff. They represent a substantial amount of erosion in the Cotswolds and probably give an indication of the age and origin of many of the misfit and dry valleys that characterise the drainage of the hills. Thus the interpretation of Phillips (1871, p. 491-2) may have some applicability:

> 'Some long interval of time undoubtedly separates us from the latest of the broad gravel-beds in the valleys of the Thames; the formation has practically ceased for thousands of years; the physical conditions of the country have changed. What were the conditions when the gravel was transported down the valleys of the Thames?
>
> Gravel is no longer accumulated, except in very small quantities, because the water-forces exerted in the valleys are unequal to transport it. The uplands are still wasted, and plenty of small calcareous stones lie on the slopes, such as might make gravel-beds; but the rain and snow are less abundant, and the floods less impetuous. The velocities of the existing currents, even in flood time, would be quite incompetent to move the larger sort of ordinary gravel, much less the large flints, larger corals, and greater masses of rock which occasionally appear in the midst of these irregular strata. To move some of these large masses would seem to require the agency of shallow ice-rafts, formed in the valley (not ice-bergs drifted from afar), because they lie not at the bottom, but in the midst of or at the top of the gravel, which could not be the case if water-streams had urged them, for then they would have been at the bottom.'

3.8 Peat

The juxtaposition of the clays and limestones across the Cotswolds gives rise to numerous calcium-rich springs and seepages, many of which have associated deposits of tufa which shall be discussed later. These springs provide a constant input of water, allowing the development and maintenance of waterlogged ground, which in some instances has led to conditions necessary for the formation of fens and fen peats. Peats have also developed within the valleys of many of the Cotswold rivers and streams and are in many instances overlain and thus obscured by alluvium (Richardson, 1933; Dury, 1958). The extent of these deposits is shown in

Figure 16. These peats, when found, can provide invaluable information on the nature of environmental changes during the Flandrian, especially in relation to changes in the vegetation on both a local and regional scale. Notable examples of these shall now be described.

3.8.1 The early Flandrian (10 000- 5000 yrs BP)

At Mingies Ditch (SP 391 059) in the valley of the River Windrush an organic lens within the floodplain sediments (horizon TS) was radiocarbon dated to 9 380±110 yr BP (Har-8366). Tree and shrub pollen frequencies were relatively high with *Betula*, *Pinus* and *Salix* the most frequently occurring types, with a few occurrences of *Quercus*, *Corylus* and *Juniperus* shown in Figure 14. Two deposits of Mesolithic age were dated 7 590±90 yr BP (Har-8354) (sample W) and 6 540±80 yr BP (Har-8355) (sample O) from Mingies Ditch. The former showed a high woodland component with high frequencies of thermophilous species including *Quercus* and *Ulmus*. *Betula* and *Pinus* still had fairly high representations in the pollen record with *Corylus* and *Salix* displaying a large expansion. The latter deposit showed abundant thermophilous woodland with *Quercus, Ulmus, Tilia, Alnus* and *Fraxinus*. *Betula* and *Pinus* were only present in very low frequencies. Shrub pollen levels had fallen as well, with only *Corylus* represented.

3.8.2 The late Flandrian (5000 yr BP to the present)

Dimbleby and Evans (1974) undertook a series of pollen and molluscan analyses from soils underlying a Neolithic barrow at Ascott-under-Wychwood, (SP 299 175), where two phases of Neolithic activity were discovered (Evans, 1972). Two radiocarbon dates from the site were obtained. One from a pit gave a date of 4 893±70 BP (BM-491), and a date of 4 735±70 yr BP (BM-492) was given by charcoal from the soil surface under the barrow (Evans, 1972). Analysis of pollen from the site showed a woodland-dominated environment with *Tilia* (lime)(14%), *Corylus* (hazel) (17%) and *Quercus* (oak), and reflected components of the pre-clearance vegetation. A possible phase of small scale clearance was shown by the pollen of Gramineae (grasses) and *Plantago lanceolata* (ribwort plantain), but it is possible that the phase could be post-barrow contamination. The phase of clearance at Ascott was abandoned as shown from the molluscan evidence (Evans and Jones, 1973).

At Buscot Lock (SU 230 980), on the Oxfordshire/Gloucestershire border, operations to create a new weir channel exposed a peat deposit in an old silted up river channel. The section contained two layers of peat with substantial logs present towards the bottom of the deposit. An *Alnus glutinosa* log from the lower layer of peat was radiocarbon-dated to 4 010±90 yr BP, suggesting a Neolithic age.

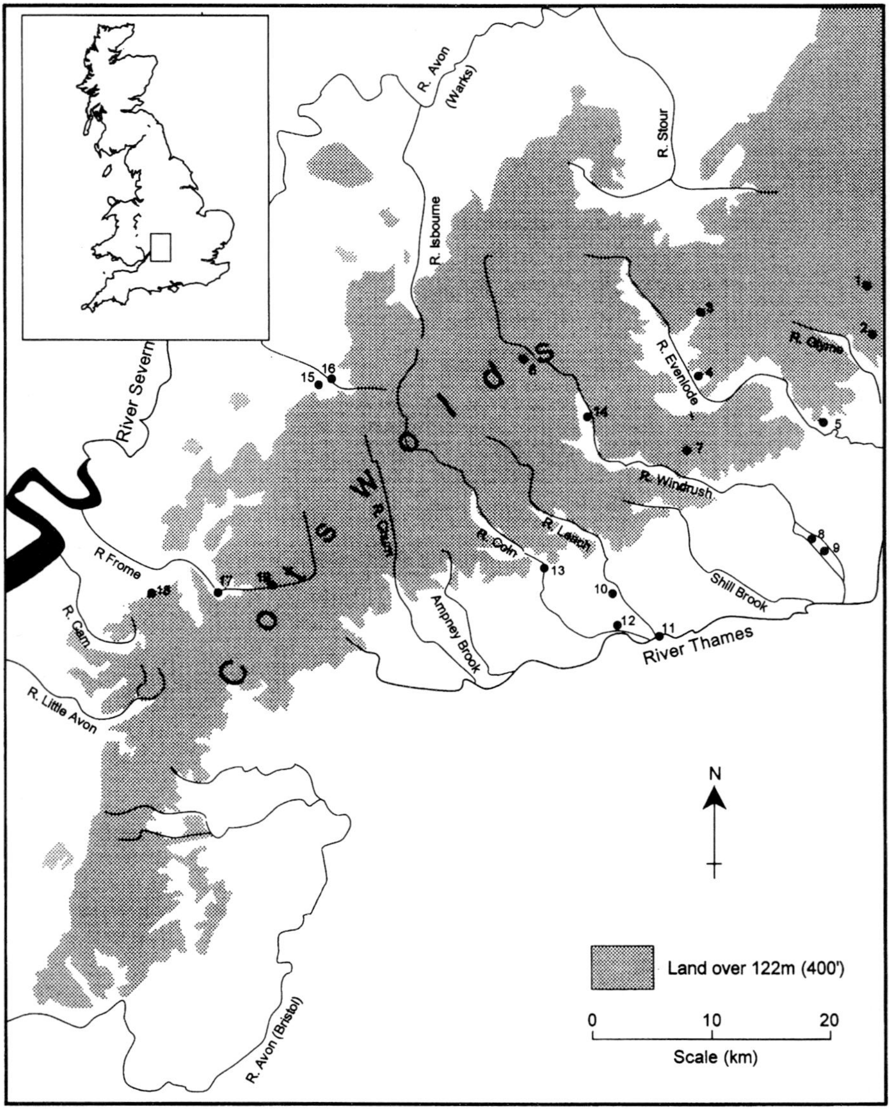

Figure 16 Location of peat deposits in the Cotswolds

List of peat sites in the Cotswolds shown in Figure 16

1. Cockley Brook
2. River Dorn
3. Salford
4. River Evenlode
5. Coombe Fen
6. Naunton
7. Taynton
8. Mingies Ditch
9. Gravelley Guy
10. Southrop
11. Buscot
12. Claydon Pike
13. Hatherop
14. Rissington
15. Leckhampton
16. Arle
17. Stroud
18. Stanley Down
19. Thrupp

Pollen from the peat was analysed by Dimbleby (Robinson, 1981). The results showed a strong alder component, perhaps along the floodplain itself, with birch, willow and hazel. *Tilia* was perhaps dominant on the limestone hills above Buscot, while oak and perhaps alder were more important in mixed woodland on the clay slopes. Other woodland species included ash and elm. Non arboreal pollen was low. Cultivation was suggested by grasses, cereals, dock/sorrel and plantain. Human activity was also attested in the vicinity by the presence of charcoal in the samples.

Gravel extraction from Gravelly Guy (SP 405 052), Stanton Harcourt, exposed a series of small organic palaeochannels. Samples from three of the channels were analysed by Rob Scaife (Unpubl.). The samples comprised dark highly humified alkaline detritus peat, containing some inorganic silt. Radiocarbon dates were obtained from two of the channels. The lowest channel, SH GG 35, gave a date of 3 740±50 yr BP (Gu-5097) (Beaker age) and the upper channel SH GG 38, 2 260±50 yr BP (Gu-5098) (late-Bronze Age). No radiocarbon date was obtained for the sample SH GG 37, which was analysed for pollen. The results of the radiocarbon determinations correspond to the two key archæological periods represented adjacent to the deposits on the gravel terrace.

The pollen from SH GG 35 (Beaker) was characterised by high values of *Alnus* (Alder, 53% TP), with *Quercus* (oak, 13% TP), and *Corylus* (hazel, 21%). Low frequencies of *Tilia* (lime), *Ulmus* (elm), *Betula* (birch) and *Fraxinus* (ash) were also recorded. Non arboreal pollen (NAP) was low, with Gramineae (grasses) and *Plantago lanceolata* (ribwort plantain) being the most abundant types. Other occurences of NAP included *Ranunculus* (buttercup), *Vicia* type (vetch), *Rumex* (dock/sorrel) with a single occurrence of a cereal grain. A few spores were also encountered, including *Dryopteris, Pteridium* and *Polypodium*.

Sample SH GG 37 was characterised by a notable increase in the number and diversity of non-arboreal pollen. *Quercus, Alnus* and *Corylus* remained the dominant woodland taxa, though reduced in number. Also present was *Fagus* (beech). The NAP was characterised by a range of species associated with arable land and pasture. The NAP included Gramineae (20%), *Plantago lanceolata* (6%), cereal type (2%), Cyperaceae (4%), *Bidens* type (daisy types) and Liguliflorae (dandelion/hawksbit). Also noted was an increase in the spore percentage of *Pteridium* (8%TP+spores).

Sample SH GG 38 showed further reductions in AP, and was dominated by Gramineae(20% TP) and Cyperaceae (19% TP), with an increase in other NAP types. This is associated with a reduction of arable taxa and an increase in wetland types.

The three samples analysed showed a progressive change in the local flora from a semi-cleared woodland/scrubby environment to a much more open one. The woodland would have been dominated by oak, lime and hazel with alder, and was cleared between the Beaker Age and late Bronze Age.

Sample 472/6 at Mingies Ditch came from a late Bronze Age tree root pit containing highly humified peat and charcoal. A date of 2 800±90 yr BP (Har-4489) was obtained from charcoal of *Alnus*. The source of the charcoal appears to be the firing of tree stumps. Pollen of *Betula, Pinus, Ulmus, Quercus* and *Alnus* were the most frequent tree and shrub types with small frequencies of *Tilia* and *Corylus*.

The earliest record of pollen analysis from an archæological site in this area is that by Dimbleby in 1958 (unpublished) from Eynsham Hall Park (SP 394 396). The site was an Iron Age fort situated on acidic drift deposits; the samples analysed for pollen were from a ditch surrounding the site. The data showed that *Quercus* (oak) woodland dominated the flora with observations of small scale clearance. An interesting point to note is the occurence of *Juglans* (walnut). This taxon was thought to have been introduced to the British Isles during the Roman occupation (Godwin, 1984; Greig, 1992). If this pollen is Iron Age in date then it is perhaps the earliest record of this species.

A radiocarbon date of 2 170±90 yr BP (Har-4485) was obtained from waterlogged twigs from a ditch (sample 10/9) which formed part of the middle Iron Age double-ditch enclosure at Mingies Ditch (Allen and Robinson, 1993). Pollen analysis by Dimbleby of peat from the bottom of the ditches suggested a relatively open landscape with AP forming 4% and shrub pollen 13% of the total sum. This was supported by coleopteran evidence. The pollen frequencies suggested species-rich grassland as represented by Gramineae (34%), *Plantago lanceolata* (6%) and Compositae liguliflorae (7%). Plant macrofossils of wet grassland or marsh species were identified and included *Ranunculus flammula, Lychnis flos-cuculi, Filipendula ulmaria, Mentha* cf. *aquatica, Pulicaria* spp. and various rush and sedge remains.

Excavations at Farmoor also revealed three Iron Age enclosures on the floodplain (Lambrick and Robinson, 1979). Waterlogged deposits yielded pollen which was analysed by Dimbleby. This inferred an open landscape similar to that at Port Meadow, Oxford, with low tree and shrub pollen (2.9%) and high Gramineæ (46%), *Plantago lanceolata* (7%) and Compositæ ligulifloræ (17%). Disturbance of the grassland in the vicinity of the farmstead was suggested by the high proportion of *Cirsium* (thistle) pollen and seeds. A radiocarbon date of 2 410±100 yr BP was obtained from one of these ditches. A later Roman enclosure was also analysed giving a radiocarbon date of 2 060±70 yr BP (Har-1926). As for the Iron Age, the pollen analysed indicated an open environment (average 3% AP and shrub). However one sample showed an increase in tree and shrub pollen, with a value of 12%.

Pollen from the Mediæval period has been recorded from a well at Claydon Pike (SP 194 001) on one of the gravel islands of the Thames near Lechlade. The pollen analysed by Greig (in Lambrick and Robinson, 1988) indicated an open grassland landscape with only 2% AP. The pollen was dominated by Gramineae (43%), *Plantago lanceolata* (7%) and Compositae Liguliflorae (24%). *Rumex, Ranunculus* and *Centuarea nigra* type were also encountered with over 3% of the total sum between them, and in keeping with a flower-rich meadowland.

3.9 Alluvium

The floodplain deposits of the modern River Thames and its tributaries commonly comprise 0.5 to 2.5m of alluvial clays overlying, with a sharp boundary, a variable thickness of sandy limestone gravels (Hazleden and Jarvis, 1979). In the Cotswolds the floodplains tend to be narrow, with the present day streams confined within steep-sided valleys cut into the limestones.

Where the rivers cross less competant strata e.g. the clays, the floodplains are broader. This is shown clearly in the valley of the River Windrush in the Vale of Bourton.

The river valleys exhibit a basically comparable sequence of Flandrian alluvial sediments with some organic and tufaceous lenses. The onset of alluvial sedimentation in the Upper Thames valley are well dated in their relationship to a series of archaeological sites and palaeochannels (Robinson and Lambrick, 1984; Lambrick and Robinson, 1988). From the third millennium BP, a major phase of environmental change occurred, leading to increased flooding, alluviation and a higher water table. This intensified in Iron Age and Roman times.

These have been correlated with accelerated forest clearance and agricultural intensification in the region with an increasing trend towards winter-sown arable land. This led to greater runoff from the hills and increased sediment loads in the drainage system (Robinson and Lambrick, 1984; Robinson and Wilson, 1987; Goudie, 1983; Parker, 1995).

3.10 Calcareous tufa deposits

A number of deposits of calcareous tufa exist in the Cotswold region. Tufa is precipitated from carbonate-rich springs issuing from the limestones within the region, and they consist of calcareous mud and silt with weakly cemented, porous and cavernous seams of calcitic material. The distribution of tufa in the Cotswolds is shown in Figure 17.

The deposition of calcium carbonate is a common feature of many freshwater environments producing a complex of varying formations (Viles and Goudie, 1990). Often associated with such formations are a range of lower plants and microorganisms, which according to Viles and Goudie (1990) are of geomorphological importance. The mechanisms of tufa formation are given in detail in Viles and Goudie (1990) and Goudie *et al.* (1993).

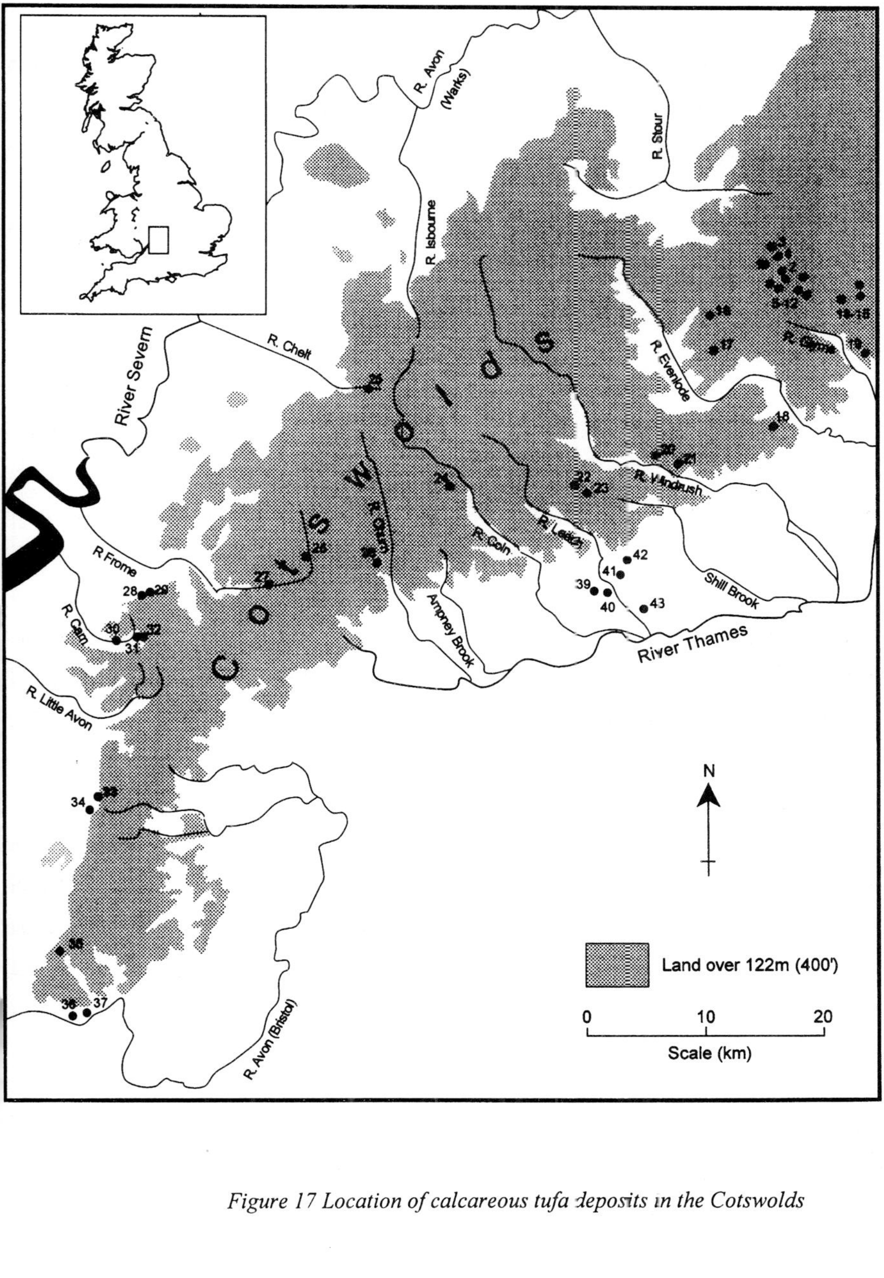

Figure 17 Location of calcareous tufa deposits in the Cotswolds

List of calcareous tufa deposits shown in Figure 17

1. Priory Mill
2. Cherwell Barn
3. Great Rollright Manor 1
4. Great Rollright Manor 2
5-12. Heythrop College complex
13. Showell Farm
14. Castle Farm
15. Little Tew
16. Sarsgrove
17. Wychwood
18. Wilcote
19. Blenheim Palace
20. Taynton
21. Swinbrook
22. Shill Brook 1
23. Shill Brook 2
24. Chedworth
25. Dowdeswell
26. Edgeworth
27. Chalford
28. Leonard Stanley 1
29. Leonard Stanley 2
30. Dursley
31. Ewelme Valley 1
32. Ewelme Valley 2
33. Birch Hill
34. Horton
35. Langridge
36. Bath
37. Warleigh Wood
38. Woodmancote
39. Great Lemhill 1
40. Great Lemhill 2
41. Common Barn Farm
42. Broughton Copse
43. Manor Farm Lechlade

Small thin spreads of argillaceous tufa occur on the valley slopes 185m and 272m to the south-west of Great Rollright Manor (SP 320 312; 321 299). At Priory Mill (SP 329 296) a dome of tufa up to 3.7m high exists. This was exploited for agricultural lime during the Second World War. At the same time an apron of tufa was worked at Cherwell Barn (SP 371 299). Willing (1986) examined two deposits of tufa in the Cotswolds for preserved mollusc and ostracod remains. Both deposits of tufa rest upon Upper Lias Clay and the source of the carbonate is primarily the Chipping Norton Limestone.

At Cherwell Barn (SP 370 299), the tufa is situated at the head of a small, deeply incised valley and occupies a roughly circular area about 200m in diameter. The sampled area was 2.9m thick and was comprised of three main lithological units. The lower unit comprised fine-grained tufa with irregular nodules and iron staining. The middle unit was marked by five bands of fine fragmented charcoal inter-stratified in the tufa. The uppermost unit consisted of hard, honeycombed tufa resting irregularly on the finer grained tufa beneath. Two radiocarbon determinations of 7 030±100 yr BP (Har-4211) and 7 230±100 yr BP (Har-4212) were obtained from charcoal layers at the site. Using the mollusc zones (Kerney *et al.*, 1980) and the radiocarbon dates, the tufa body appears to have formed between 8 500 yr BP and 6 000 yr BP. Throughout the deposit the presence of predominantly shaded or humid species of snails was apparent with occurrences of species such as *Acanthinula aculeata* and *Carychium tridentatum*.

Few aquatic snail species were thus found, suggesting the presence of swampy ground with a few small pools rather than open water.

At Priory Mill (SP 333 295) tufa occupies the head of a small narrow valley which is the source of the River Stour. The tufa is 2.5m thick and overlies 0.5m of organic silt. The basal organic sediment gave a radiocarbon date of 8 720±100 yr BP (Har-2675) (Willing, 1986). The molluscan record showed a similar picture to that at Cherwell Barn.

There are two spreads of tufa on the southern slopes of the valleys west of Little Tew. One of the deposits is around 1.5m thick (SP 362 288) and is situated near Showell Farm; the other (SP 309 285), north-east of Castle Farm, was worked to a depth of 1.5m during the Second World War.

Small aprons of tufa occur on both sides of the valleys to the north and west of Heythrop College (SP 344 272; 270 352; 357 267; 362 258; 359 275). Most have been formed by springs thrown out from the top of the Upper Lias, but a small spread (SP 372 259) south of Kitewood has been produced by springs from near the base of the Middle Lias. Some of these spreads display a micro-topography of tufa dams and pools.

In Sarsgrove Wood (SP 307 245) a tufa fan lies in a small tributary valley of the Sars Brook. The brook itself flows over a bed entirely consisting of tufa. Arkell (1947a) mentions that in the grounds of Blenheim Palace running along the east of the park exists a dry valley (SP 438 187) 'half-choked' with tufa. Between North Leigh and Wilcote, near Bridewell Farm lies a deposit of tufa 0.5km long (SP 382 147). Arkell (1947a) mentions that this deposit is at least 2.5m thick and floors the valley.

In the valley of the Shill Brook at Burford Signet, two areas of tufa each around 1km long have formed from the precipitation of carbonate-rich springs issuing from the base of the White Limestone (Beckinsale, 1982a).

During the recent re-mapping of the Cirencester sheet the British Geological Survey have noted a number of calcareous tufa deposits (Sumbler, pers. comm.). Five deposits were mapped in the vicinity of Southrop and Lechlade (Horton, 1988; Sumbler *et al.*, in prep), which lie near the

Cornbrash-Kellaway Clay boundary which gives rise to strong springs that were previosly thrown out at this level. The largest deposit occurs in the floor of the R. Leach valley, north-east of Great Lemhill Farm (SP 204 018), in which a mound up to 1.5m has developed. Smaller tufa deposits were noted north of Common Barn Farm (SP 219 024), east of Broughton Copse (SP 231 032), an active spring at Great Lemhill (SP 0235 0182) and east of Manor Farm, Lechlade (SP 227 003).

In the valley of the R. Windrush, the Coombe Brook tributary has a number of tufa deposits which have developed from springs issuing from the base of the Clypeus Grit (Sumbler, 1994; Williamson, 1995). Williamson (1995) notes that

> "Calcareous tufa forms small build-ups and barrages.....in the wooded valleys of the Coombe Brook. The chance presence of a substrate of woody material, leaves and talus appear to be a pre-requisite for initiating the more significant precipitates seen."

Sumbler (1994, 1995, pers. comm.) has noted that the alluvium within the valleys of the Rivers Windrush and Leach contain lenses of calcareous tufa locally. He also noted in the winterborne of the Ampney Brook that the alluvium contains scattered pellets of tufaceous carbonates.

Richardson (1904, p.154) noted 'a considerable deposit of 'travertine' was forming at the north side of Moor Wood near Combend where a stream drains off the Fuller's Earth.

Murchison (1844, p.62) describes what is now termed calcareous tufa from the headwaters of the the River Chelt.

> "The Chelt, which gushes out upon the upper clay of the Lias formation, after having percolated the calcareous strata of the Inferior Oolite...is charged with a certain quantity of carbonate of lime held in solution, which it deposits in the early part of its course upon any substances over which it happens to flow. This process of incrustation may be observed in actual operation, on a small scale, below Sandywell Park, at the source of this river. The quantity of carbonate of lime held in solution is small, and upon escape of the carbonic acid gas, the earthy residuum is only for a short space left adhering to and covering the contiguous moss or branches. To these, which have been called petrifying springs, the term encrusting is more correctly applicable."

Playne (1874) described tufa at Chalford, in the valley of the R. Frome (or Stroudwater). Large quantities were removed during the construction of the GWR line in 1842. The tufa has been used for ornamental purposes in gardens and fernaries and also used for building purposes.

Calcareous tufa and petrifying springs have also been noted at Hen Wood, Edgeworth in the valley of the River Frome (or R. Stroud) by Duncan and Richardson (1924), where springs are thrown out by the Fuller's Earth from the base of the Great Oolite (Richardson, 1933, p.238).

At Dursley, in the valley of the River Cam (or R.Ewelme), Richardson (1927a,b) described the tufa deposits which have formed from springs issuing from between the Middle Lias and Marlstone Rock Bed. The Dursley tufa has in the past been quarried and used for building purposes, and was used for filling in vaultings of both Gloucester and Worcester Cathedrals, and also those at Kenilworth Castle, due to its lightness and durability, and also in the walls of Berkeley Castle, the chapter house at Wenlock and at Glastonbury (Playne, 1874; Anon, 1925; Richardson, 1927a).

Cave (1977) has noted three small deposits of calcareous tufa in the Cotwolds whilst mapping the geology of the Malmesbury district. At Horton (ST 766 850) springs thrown out in the Cotteswold Sands have precipitated tufa, which is up to 1.5m thick consisting of cream-coloured, calcareous silt containing gastropod shells. Two smaller deposits have formed in the Ewelme Valley near Wresden Farm where springs have been thrown out by clay beds in the Dyrham Dilts.

The BGS sheet 234 (Gloucester) shows two deposits of calcareous tufa at the base of the escarpment 1km south of Leonard Stanley (ST 797 028 and 799 027) at the junction between the Dyrham Silts and Lower Lias. The larger of the two covers approximately 200m x 200m.

Butler (1983, p.109) augered sections in the drifts of the small coombe complex at Birch Hill (ST 767 860) in the south Cotswolds. This revealed tufaceous deposits beneath the plough soil, which was full of carbonised and calcified twigs and leaves and also yielded a number of molluscs. Though now dry the former springs from which the tufa was deposited, would have issued from the Cotteswold Sands. Findlay (1976, p.18) noted tufa deposition near Langridge (ST 733 692), where

> "..slopes are flushed with...secondary deposits of lime".

He also noted tufa at Warleigh Wood (ST 787 628) near Bath, at the base of landslipped slopes.

In the extreme south of the Cotswolds, calcareous deposits, especially aragonite, have been deposited by the thermal waters of Bath Spa (ST 753 648) (Rastall, 1926). The origin and composition of the Bath thermal waters have been the subject of much scientific curiosity, and the results of systematic investigation of these hot springs may be found in Kellaway (1991).

Tufa is still forming in the heads of many of the small streams fed by the springs in the Cotswolds, for example at Chedworth, Swinbrook, Taynton and in Wychwood (personal observations) where snails, stones, bottles and stems of rushes and reeds, twigs and mosses have become encrusted with tufa. However, as elsewhere in Europe it is likely that tufa deposition is less extensive now than it was in the early and Middle Holocene (Flandrian) (Goudie *et al.*, 1993).

3.11 Cheltenham Sand

In and around Cheltenham, there are thick deposits of quartzose sand with occassional Jurassic pebble debris, known as the Cheltenham Sand (Green, 1992). It is medium grained, well sorted, and quartzose with rounded grains and contains small proportions of calcium carbonate, many as ooliths or small angular limestone fragments (Briggs, 1975). The Cheltenham Sand was apparently wind-derived (Richardson, 1912, 1913a; Tomlinson, 1940; Briggs, 1975) from the deflation of the Severn Terraces backed up against the escarpment by westerly winds, and fluvially re-distributed in the late Devensian. Within the Cheltenham Sands remains of mammoth have been found at Charlton Kings and from the old railway cutting between Charlton Kings and Leckhampton (Richardson and Sandford, 1963b). Richardson and Sandford (1963b, p.171) reported unpublished accounts by Buckman, Calloway and Richardson for a high altitude site of Cheltenham Sands at approx 120m in the old railway cutting on the Cheltenham and Kingham line due south of the Dowdeswell reservoir, near Charlton Kings (SO 987 196).

Chapter 4

4.1 The valley systems

W.M. Davis wrote some of his classic geomorphological papers after visits to the Cotswolds (Davis, 1899; 1909) and the area has subsequently been studied by many other workers. They have been mainly concerned with the drainage of the region, especially the deeply incised rivers (Buckman, 1899, 1900; Beckinsale and Smith, 1955; Dury, 1955, 1965; Beckinsale, 1970).

The Cotswolds are drained by a large number of streams and rivers (Figure 18), the largest of which is the River Thames. Within the Upper Thames basin, 'consequent' dip-slope rivers and 'subsequent' strike streams dominate the drainage and relief (Goudie, 1976). The headwaters of the River Thames rise at Thameshead in Gloucestershire (Richardson, 1923b) and flow through the region a little north of eastwards. They cross onto the Oxford Clay Vale and near Appleton the river turns northwards and curves around Eynsham, Wolvercote and Oxford, leaving the Upper Thames valley in the south-eastern corner at Goring (Richardson *et al.*, 1946). Until reaching Wolvercote, the Thames is a strike stream, meandering along the Oxford Clay outcrop, but as it passes onto the overlying Corallian beds 5 km south of Oxford, becomes a dip-stream where its valley narrows considerably.

Except for the Cole and Ock no major tributary enters the Thames on its right-hand bank, but on its left hand bank it is joined by numerous rivers and minor streams. The most notable tributaries of the Thames are the Rivers: Churn (entering at Cricklade), Ray (entering at Cricklade), Coln (entering at Inglesham), Cole (entering at Inglesham), Leach (entering at Lechlade), Windrush (entering at Newbridge), Evenlode (entering at Cassington), Cherwell (entering at Oxford), Ock (entering at Abingdon) and Thame (entering at Dorchester). Details of many of the tributary streams and rivers can be found in Richardson (1923a, b, 1924, 1925b, 1926a, b, 1927a, b, 1929, 1930, 1933). The Thames has a broad alluvial floodplain mainly comprised of clays and clay-loams. The river still regularly reaches its overbank stage and floods. Human exploitation and control of the drainage basin has in part influenced the regional hydrology (Robinson and Lambrick, 1984).

To the north of the Cotswold scarp near Rollright, the River Stour drains the northern part of the Vale of Moreton and the margin of the east facing slope of the north Cotswolds into the Warwickshire Avon, in the Vale of Red Horse (Richardson, 1929; Beckinsale and Beckinsale, 1980). One tributary has made a far-reaching incision at Dovedale, near Blockley.

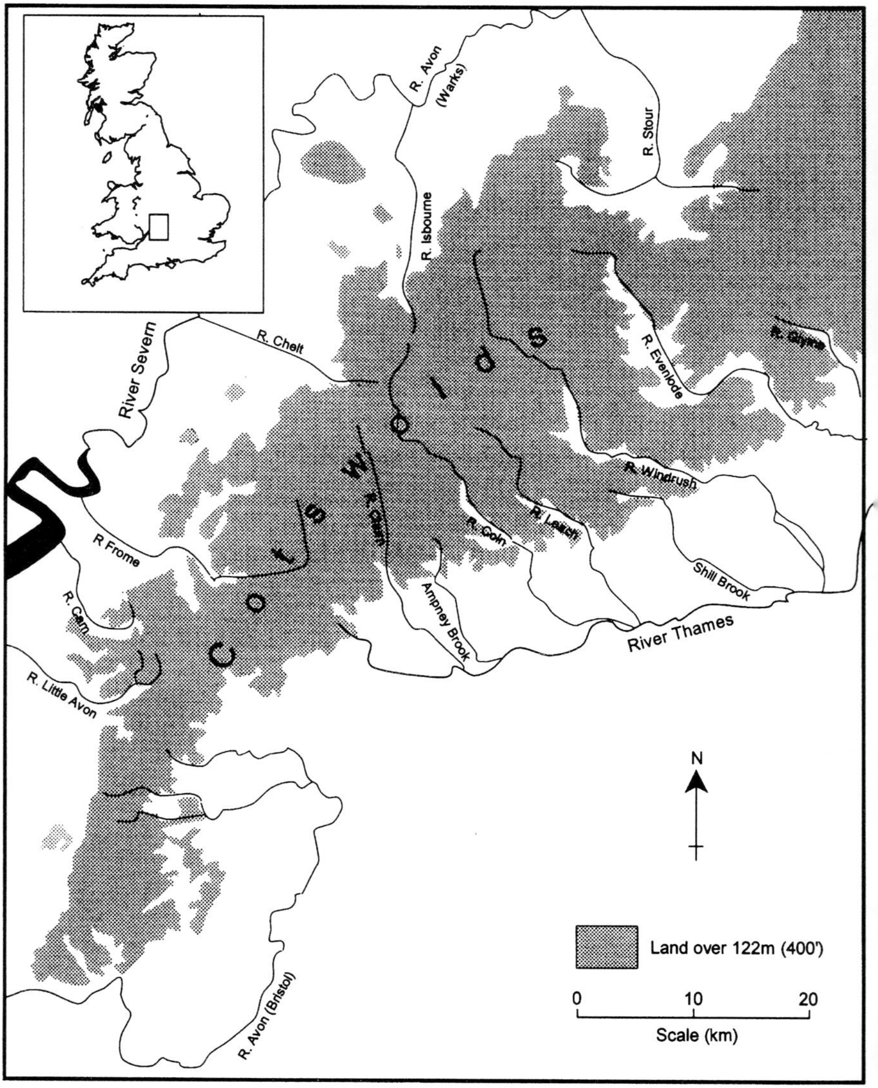

Figure 18 Rivers and streams of the Cotswolds

The River Isbourne has eroded the Winchcombe embayment and drains the vale at the north facing scarp of the main hill mass of the north Cotswolds, the northern part of the Charlton Abbots Valley, *via* the Babesmore Brook, and the north eastern flank of the Cleeve Hill plateau. This river has developed along the axis of the Winchcombe anticline. Thus this structure would appear to have facilitated erosion by the River Isbourne. The main north Cotswolds escarpment and the outliers of the Woolstone-Oxenton, Alderton-Dumbleton and Bredon Hills are separated by the westward flowing Carrant Brook and its tributaries. The River Chelt drains part of the southern portion of the Cleeve Hill Plateau and the vale on the western side of the Cotswold escarpment.

In the southern Cotswolds dissection of the high crest of the main rock mass by the Bristol Avon, the Little Avon, Cam Brook and tributaries of the River Frome give rise to steep valleys. Both the Bristol Avon and River Frome start on the dip slope of the Cotswolds, then curiously they turn westwards, eroding deep valleys through the scarp slope. The Painswick Brook and River Cam have incised back almost to the scarp edge (Figure 18) (Murray and Hawkins, 1973).

The main dip slope rivers occupy wide deep-set valleys cut down to Lias clays at or near their sources, but enter upon 'gorges' incised into the oolitic limestone.

4.2 Long profiles of Cotswold streams

Most rivers have concave long profiles in which there is a systematic descrease of gradient in a down valley direction. Breaks in the regular curve may be caused either by structural or lithological controls or by the development of knickpoints caused by rejuvenation (Beckinsale and Smith, 1955). Rejuvenation itself may be caused by tilting, uplift, or changes in discharge characteristics and the balance between stream erosion and deposition. Streams and rivers adjust towards a graded state. This is established first in the downstream segments of rivers, and gradually extends upstream.

Beckinsale and Smith (1955) produced longitudinal profiles of the Windrush and its main tributaries (Figure 19). Other long profiles have been constructed for the scarp face (Figure 20) and dip slope valleys (Figure 21). What is evident from these figures is that the long-profiles, though they have some irregularities, do display the normal concave long-profiles, and that the upper portions of the scarp face valleys are, as one might expect, particularly steep. This is shown by the relationship between the graded river profiles and the exponential curve for each river valley. The exponential curves represent sharply concave theoretical profiles of least work. Thus a theoretical profile for uniformly distributed work should have a constant slope. Grade is a condition. Thus, the closer the river or stream is to being graded the nearer the relationship displayed between the long-profile for each river or stream and the theoretical exponential relationship is to -1. For example the River Thames from its watershed to Witney is in a near graded state. The rivers of the Cotswolds tend to show graded profiles with values ranging between (-0.96 to -0.99) for the dip slope rivers and (-0.95 to -1) for the scarp slope rivers and streams. The tributary streams for both scarp and dip slope streams also tend to be less graded than the larger streams and rivers which they feed. Larger rivers attain grade earlier than smaller streams. A trunk stream may be at grade whilst its smaller tributaries are still grading headward into undissected slopes.

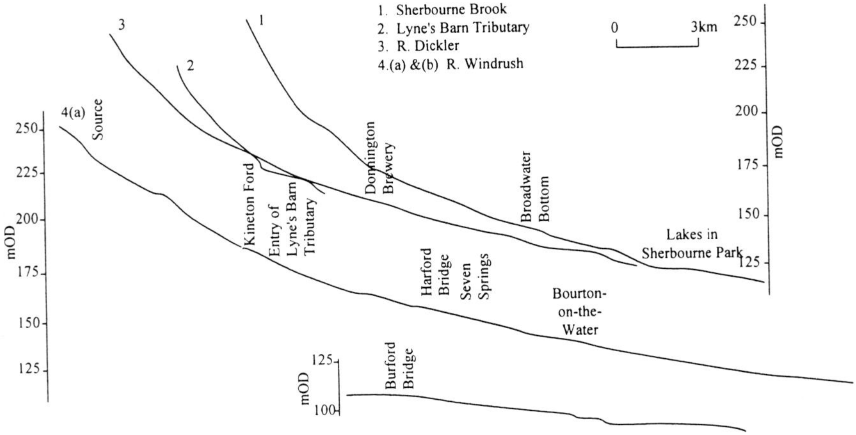

Figure 19 Longitudinal profiles of the valley of the R. Windrush and its main tributaries (after Beckinsale and Smith, 1955, fig. 1)

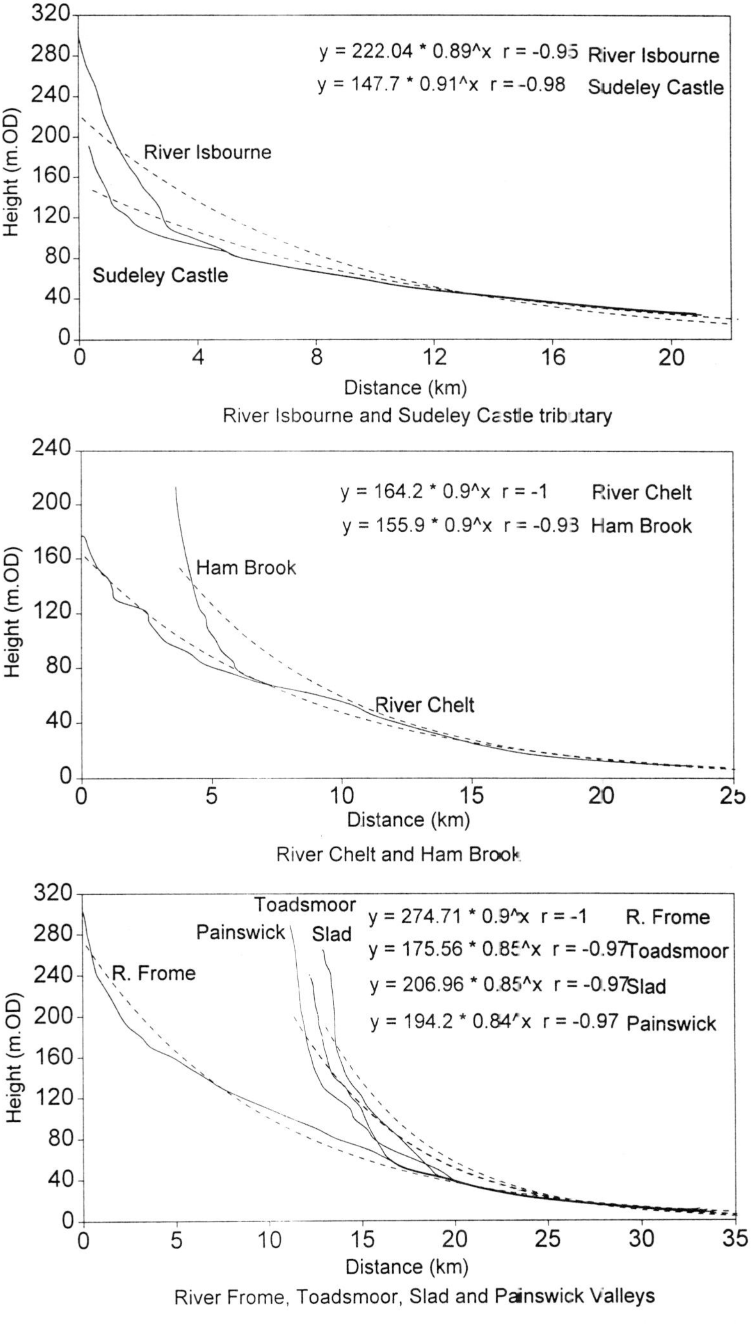

Figure 20 Longitudinal profiles of the scarp face rivers

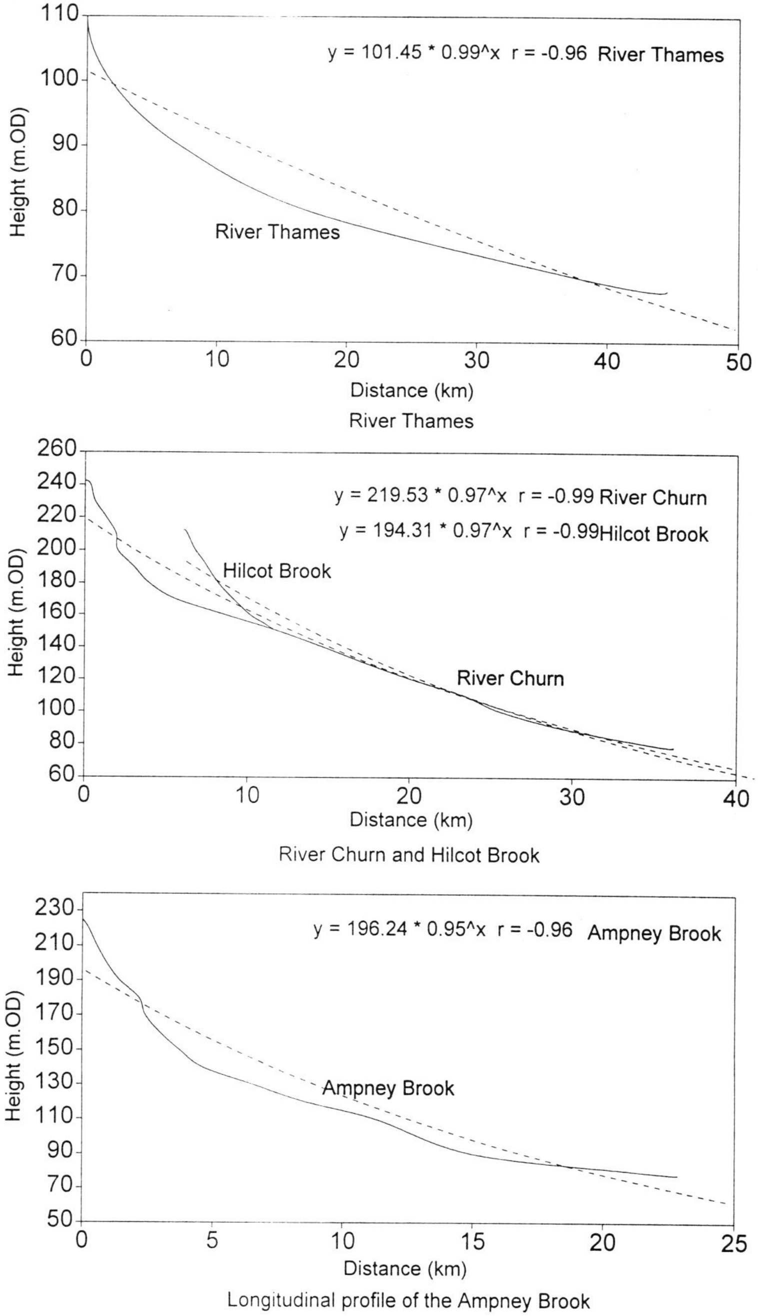

Figure 21 Longitudinal profiles of the dip slope rivers

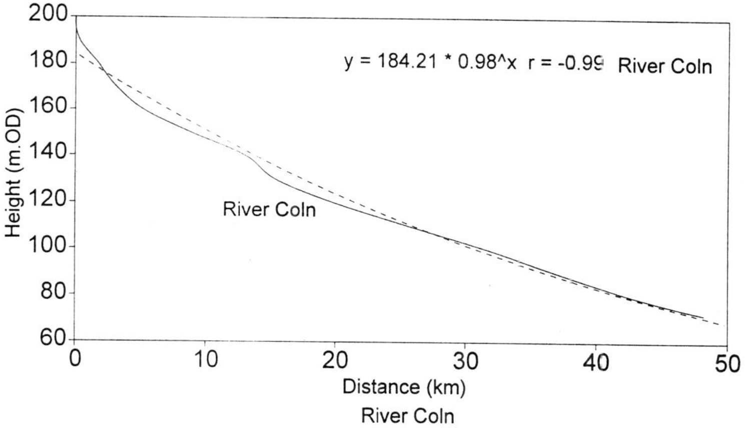

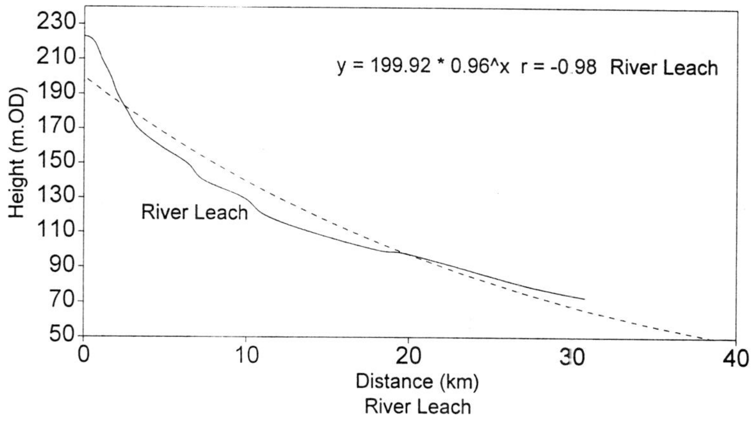

Figure 21 (continued) Longitudinal profiles of the dip slope rivers

4.3 Morphometry-the nature of the valley nets

The nature of the valley nets in the Cotswolds was ascertained from morphometric analyses of 1: 25 000 O.S. maps (Goudie and Hart, 1975). Data from this work are shown in Table 5. The analyses were restricted to third order stream basins using Strahler's method (Strahler, 1952), and for the construction of valley networks lines were drawn through the crenulations of the contours. This method has also been adopted by Morgan (1971), Paterson (1970) and Goudie *et al.* (1980). The results show that there are no statistically significant differences between the bifurcation ratios, drainage densities, channel frequencies and Melton infilling indices between the different rock types within the area (de Courcy Wheeler, 1974). Thus in essence the valley nets appear to be uniform through the area with respect to these variables.

Table 5 Miscellaneous basin morphometric characteristics for the Oxford region (from de Courcy Wheeler, 1974)

	Chalk	Corallian	Oolite	Clay
Bifurcation Ratio	3.34	3.32	3.77	3.32
Mean stream lengths (km)				
order 1	5.10	4.69	4.35	3.62
order 2	3.16	2.83	1.79	2.03
order 3	1.79	2.48	1.27	1.90
Channel frequency	1.80	1.25	1.92	1.65
Melton's basin infilling index	0.16	0.18	0.20	0.19
Valley density (km/km^2)	0.96	1.00	0.88	0.85

The great bulk of the valley network is probably a relict Pleistocene feature which was imposed by the formerly less marked permeability differences between rock types produced by permafrost cover.

In essence, the valley densities of the Cotswolds are low in comparison with those of many sedimentary rocks elsewhere in England (Table 6) though densities in lowland England tend to cluster around 2 to 3 km km^{-2}.

Table 6 Morphometric data for small English catchments (after Goudie and Hart, 1975)

Source	Location	Mean bifurcation ratio	Mean valley density (km/km^2)
Morgan, 1971	English Chalk	4.0	-
Brown, 1969	Chilterns	-	2.66-3.84
Morgan, 1971	Weald Clay	-	2.3
Morgan, 1971	Ashdown Sand	-	2.0
Morgan, 1971	Tunbridge Wells Sand	-	1.7
Gregory, 1966	Otter Basin	-	2.14
Gregory, 1966 (5 basins)	South Devon	-	3.03
Brunsden, 1968	Dartmoor Granite	3.5-3.9	1.61-2.24
Brunsden, 1968	Middle Devonian	3.8-4.2	2.92-2.9
Brunsden, 1968	Lower Devonian	3.9	2.86
Brunsden, 1968	Culm Measures	3.4-3.7	3.23-4.78
Chorley & Morgan, 1962	Dartmoor	-	2.14

4.4 Dry valleys and misfit streams

The underfit streams fed by spring sources, occupy the principal valleys, while numerous dry valleys, which are seldom, if ever, occupied by a stream channel, are cut into the broad upland surfaces (Paterson, 1970). The dry valleys are often tributaries to the misfits, and the origins of the two types of drainage system are almost certainly closely connected.

Misfit streams are streams which are much smaller (by a ratio of around 10:1) than the valleys in which they flow in respect of width and meander wavelength. The valleys may contain a considerable depth of fill as shown in Figure 22 a (Dury, 1965). Valley-in-valley forms (Figure 22 b) also occur along some of the Cotswold headstreams especially the Windrush, and Buckman (1899, 1900) believed that they resulted from the progressive truncation of headstreams as the Severn-Avon system developed. Certainly, fluvial processes under conditions similar to the present day are inadequate to account for either the character or the extent of the misfit streams and dry valleys, and many investigators now propose that the prime cause of their development has been long term climatic change (Beckinsale, 1954, 1970). Dury (1965) suggested that the abandonment of the largest channels occured in the Late Glacial, and showed that for the River Dorn in Oxfordshire channel infill occurred during the early Flandrian around 9000 yr BP.

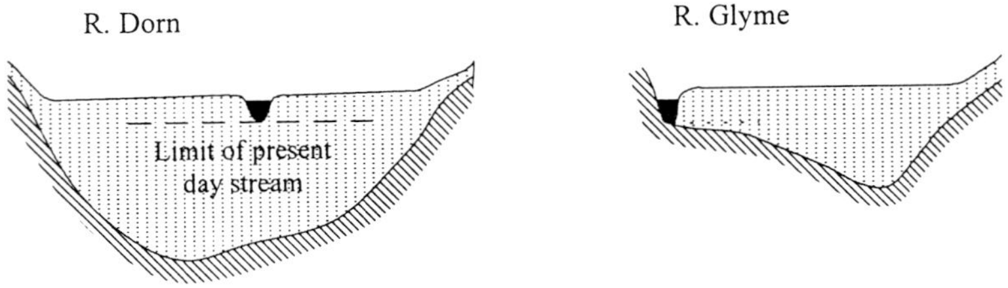

Figure 22a Diagram showing the contrast between underfit streams and bedrock channel size (after Dury, 1965)

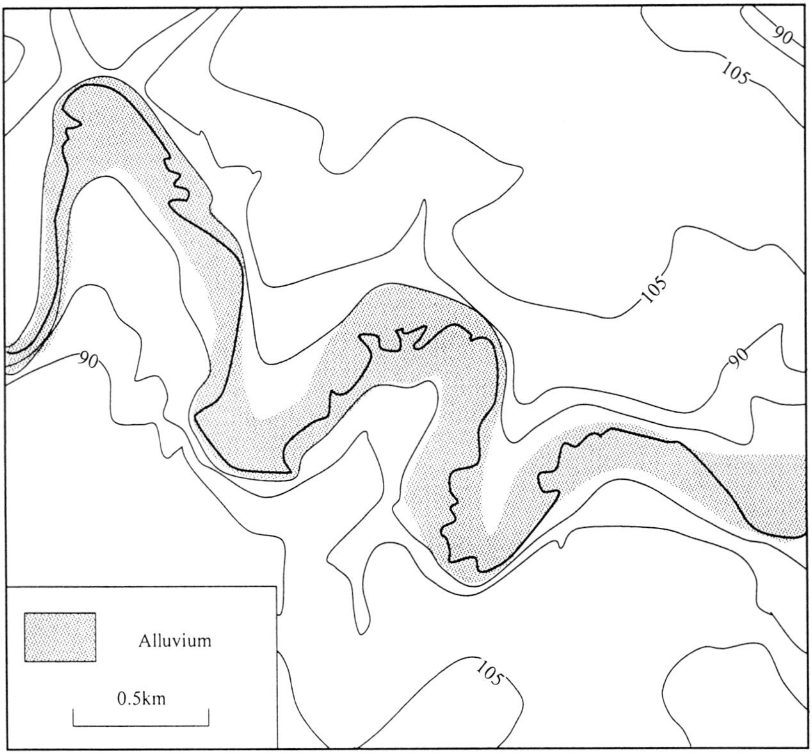

Figure 22b Valley-in-valley forms of the R. Evenlode (after Small, 1970)

4.5 Valley meanders and abandoned meanders

The river valleys in the Coswolds exhibit a number of features known as valley meanders in which the relationship between stream and valley size is most impressive where the valley itself describes regular and symmetrical meanders. These are meanders which are cut in bedrock which have a greater wavelength than that of the contemporary river pattern. It has been suggested that they were formed in softer, overlying clays and that their impression was then left in the more resistant, underlying strata (Beckinsale, 1954, 1970; Beckinsale and Beckinsale, 1980). Such meanders are well developed in the Cotswolds river valleys (Figure 23). Beckinsale (1970) pointed out that the valley meanders of the Windrush below Burford were probably superimposed from a surface cover of Oxford Clay and Forest Marble. He also suggested that the Forest Marble in the Leach Valley had only been breached in fairly recent geological time. Perhaps the best known examples, however, are the valley meanders of the River Evenlode (Arkell, 1947 a, b) in the lower part of the so called 'Evenlode gorge'. These features have drawn the attention of a number of workers over the last century (Buckman, 1899, 1900; Dury, 1955).

Davis (1899) was especially fascinated by the misfit nature of the Cotswold rivers and referred to

> 'the striking discordance between the minute irregularities of the stream and the swinging curvatures of its valley'. (p.88)

He believed the valleys had been originally moulded by the

> 'vigorous, able-bodied rivers'

whereas at the present time a

> 'little staggering stream like the Cherwell...wanders about, as if bewildered, on the flat valley floor.'

He recognised that misfitness could perhaps result from glacial overflow (Davis, 1909, p. 151), or from

> 'some climatic change of external and obscure origin' (1899, p. 93),

and from

> 'increased evaporation following the destruction of ancient forests and the cultivation of the ground' (1899, p.93).

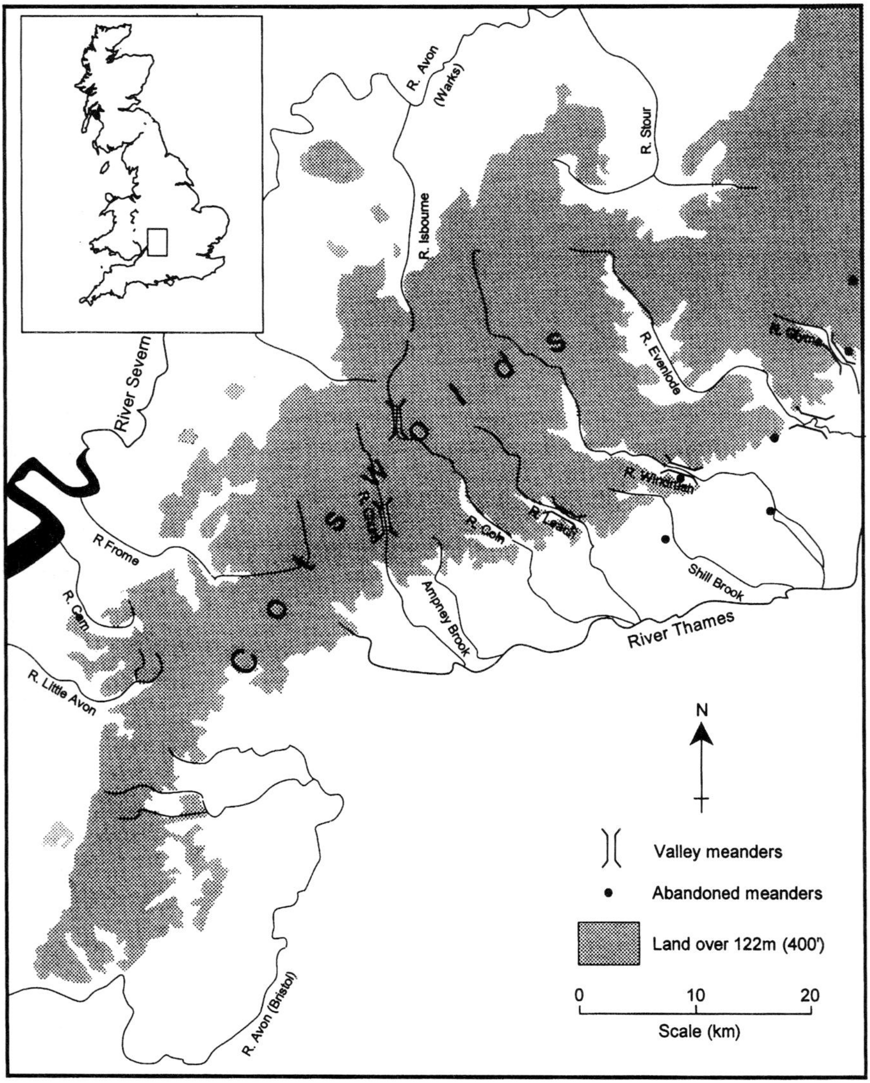

Figure 23 The location of valley meanders and abandoned meanders meanders meanders in the Cotswolds

However, his prime mechanism was the role of the scarp retreat and drainage pattern evolution, whereby the Cotswold dip-slope streams were seen as beheaded consequent streams. He remarked that

> 'of all the river contexts in England, that by which the Thames system has been shorn of its original importance is the most interesting' (Davis, 1909, p. 144).

Dury (1958, 1965) has shown that valley meanders were produced during periods of higher run-off and higher peak discharges before stream shrinkage which led to contemporary underfit streams. He argues that such discharges can only have occurred during the Quaternary, when factors such as higher rainfall, reduced percolation and lower evapotranspiration caused far greater runoff than that of today.

Most valley meanders are occupied by the small, misfit modern streams. During the Flandrian, the channels have become choked with alluvium, and the meandering valleys have become occupied by underfit streams. However, a number of streams have abandoned meanders and shortened their courses by cutting across the neck of land between the meander loops. There are a number of good examples in the Cotswolds which shall now be discussed. In the valley of the R. Windrush Sumbler *et al.* (in prep) note an abandoned meander at Bourton-on-the-Water (SP 174 194) infilled with older alluvium 0.5m above the floodplain. At Asthall a large, classic shaped abandoned meander is to be found. The now dry valley is floored with gravel and older alluvium. In addition to the Asthall meander, Richardson *et al.* (1946, p.6) suggested that there was another at Witney

> "...where a considerable part of the town seems to stand on a low meander-core of Cornbrash."

In the valley of the R. Evenlode, a small tributary brook between North Leigh and Ashford Mill has been assigned as an old meander course (Richardson *et al.*, 1946, Arkell, 1947a). This feature has been the focus of much debate (Bromehead, 1909; Davies, 1923). Davies (1923, p.90-91) assigned it to the Evenlode, whilst Bromehead (1909, p.141) regarded it as part of an old course of the Windrush. Davies also noted that the crescentic sweep of the steep valley side north of Long Hanborough was connected with an abandoned meander of the Evenlode. The BGS Witney sheet 236 confirms this, with alluvial material filling the abandoned channel course. On the west bank of the R. Glyme, a tributary of the R. Evenlode, there is an abandoned meander at Lesure Hill to the north of Woodstock, which is relatively small.

The Tackley meander of the R. Cherwell occurs to the north of Enslow Bridge with Pound Hill at its core. This meander was described by Davies (1923, p.92-94). However, Richardson *et al.* (1946, p.7) noted that the gravel within the abandoned meander loop was not in the floor of the meander valley as thought by Davies, but higher up on the south-western side of the curve.

Richardson *et al.* (1946, p.8) noted an abandoned meander in the valley of the Shill Brook, a small intermitent Cotswold stream between the Rivers Coln and Windrush. Beckinsale (1982a, p.74) describes this meander in some detail and describes this feature south of Kenn's Farm, near Carterton, as

> 'a distinct abandoned valley-meander around a small knoll or hillock of Cornbrash'.

There is no common consensus as to the causes of valley meanders, misfit streams and dry valley development.

Table 7 Hypotheses of dry valley formation (after Goudie and Hart, 1975)

Uniformitarian

1. Superimposition from an impermeable cover of Tertiary Beds, Plateau Drift or Fullers Earth
2. Joint enlargement by solution through time
3. Cutting down of major through-flowing stream
4. The Chandler-Fagg hypothesis of scarp retreat
5. Cavern collapse
6. River Capture (e.g. of Thames tributaries by Severn/Avon tributaries)
7. Rare events of extreme magnitude (e.g. extreme floods)

Marine

8. Non-adjustment of stream to a falling Pleistocene sea-level and associated fall of ground-water levels
9. Tidal palaeomorphs

Palaeoclimatic

10. Overflow from proglacial lakes (such as Lake Harrison)
11. Glacial scour, possibly during the Anglian
12. Glacial meltwater
13. Higher rainfall and/or reduced evaporation
14. Spring snow melt under periglacial conditions
15. Runoff from impermeable permafrost

Table 7 shows the main categories which have been postulated (Goudie and Hart, 1975; Goudie *et al.*, 1980). These have been categorised into three main themes. First are uniformitarian hypotheses, which require no major changes of climate or base level but merely the operation of normal geomorphic processes through time. Second, are marine hyptheses, which are related to changes in base level, and third are palaeoclimatic hypotheses which are primarily associated with the climatic changes of the Quaternary. Goudie *et al.* (1980) suggest that some of the hypothseses have broad applicability throughout the region e.g. 2, 7, 8, 13, 14, 15 whilst others may account for individual cases e.g. 5, 6, 10, 11. It is not yet possible to suggest which hypotheses are likely to have been the most important. However, solifluction deposits and terrace spreads suggest that the cold phases of the Pleistocene did see intense fluvial erosion, frost weathering and mass movement (Goudie *et al.*, 1980).

Beckinsale (1954, 1970) has carefully argued and put forward a strong case for the hypotheses of erosion by niveo-fluvial processes operating in the oversize valleys of the Cotswolds. He lays great stress on the regime of the streams that would have existed during periglacial conditions - a 'nival' regime, when great valley erosion would occur. Terrace aggradations during the cold periods of the Quaternary are in accord with this theory (Goudie and Hart, 1975; Briggs *et al.*, 1985), based on the idea that, under periglacial conditions, material was eroded from the Cotswolds and deposited in the Thames valley as terraces. In the southern Cotswolds, Kellaway *et al.* (1971) have suggested that the drainage system may have been influenced by early glaciation, but their views on the extent of glaciation in southern England have received relatively little support.

4.6 Wind gaps and water gaps

Small (1970, p.78) discusses the nature of water gaps and has noted that

> "The continuity of most cuestas is broken at some point by large through valleys occupied by permanent streams. The latter are usually, though not always major, dip-slope consequents which for one reason or another have resisted encroachment by neighbouring consequent systems or have had their own catchments ahead of the escarpment, through subsequent stream growth or river capture".

The Cotwolds in general lack water gaps, except for the Bristol Avon, which cuts, unusually, against the dip slope of the southern Cotswolds at Bath. This is in contrast with many other escarpments in England e.g. the North York Moors, where a number of water gaps are to be found.

Wind-gaps mark through valleys, once occupied by sizeable consequent streams, which have since been beheaded by subsequent streams developing ahead of the escarpment. They are now dry in their uppermost parts, but may contain small misfit streams in their lower courses. In the Cotswolds an interesting sequence of wind-gaps exist (Figure 24).

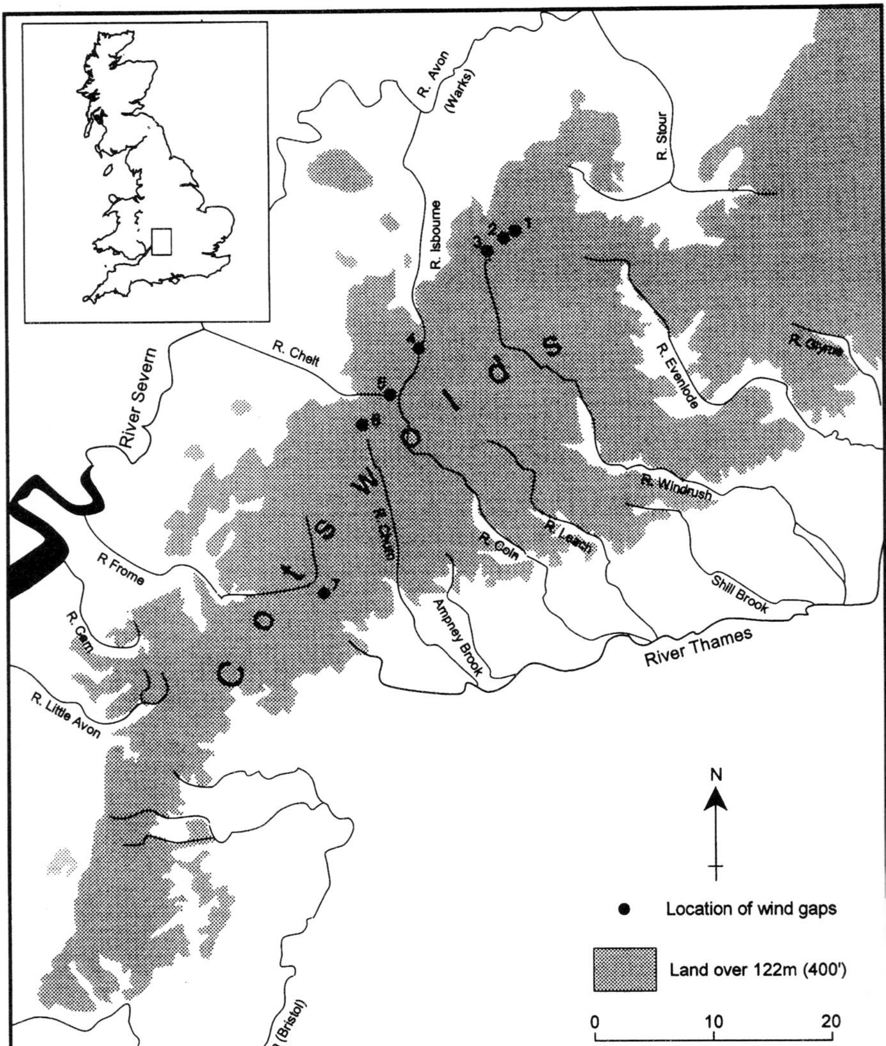

Figure 24 Location of wind gaps in the Cotswolds

Wind-gaps occur at the head of four main Cotswold rivers, namely, the Windrush, Coln, Churn and Frome and are as follows:

1. Oat Hill (SP 093 333) at the head of the Windrush valley
2. Stumps Cross (SP 082 298)
3. Lyne's Barn Farm (SP 062 278) at the head of a tributary valley of the Windrush
4. Charlton Abbots (SP 035 237) at the head of the Coln valley
5. Andoversford (SP 009 205) at the head of a tributary of the Coln
6. Seven Springs (SO 968 177) at the head of the Churn valley
7. Sapperton Wind Gap (SO 973 023) between the Frome and Thameshead

Interpretation of the wind gaps as resulting from a sequence of stream development on gently dipping strata was put forward by Buckman (1899, 1900). This traditional view of the Cotswold wind gaps, as indicating the former extension of the drainage pattern to the north-west has been summarised by Beckinsale (1954, 1970) and Beckinsale and Smith (1955).

4.7 Karstic forms and processes

The limestone of the Cotswolds are not well known for the development of classic karstic terrain. Closed depressions and caverns are rare, though fluvio-karst in the form of dry valleys is widespread. Sweeting (1972) has stated that the essential characteristic of areas of karst landforms is the presence of vertical and underground drainage, in which the drainage of surface streams is affected by dolines and swallow holes. Thus, the main difference between karstic and non-karstic areas are that in the former, drainage in limestone sinks into the ground and does not become integrated into a dendritic system of surface rivers. In the latter, surface waters become organised and systematised into valleys and connected networks. Figure 25 shows the locations of some of the karstic phenomena in the area.

There are some small karstic forms in the Cotswolds. Arkell (1947, p.246) and Richardson *et al.* (1946, p.9) refer to some 'swilly holes' in a wood above Waterman's Lodge, Wychwood, at the base of the Forest Marble Clay. At Coombe and Ramsden Heath swilly holes have developed at the base of the Oxford Clay. Richardson *et al.* (1946) and Beckinsale (1982b) have noted swilly holes in three places near Sturt Farm and also in the valley of the Shill Brook downstream from Signet, near Burford, in which water from this stream has been observed disappearing down.

Richardson (1913b) described a swallow hole around ¾ km east of Seven Springs. This swallow hole is a triangle-shaped depression about 15m long, 4-5m deep, narrowing downwards so that the lower portion is circular with a 2m diameter. The hollow is a situated near an outlier of Forest Marble Clay and water disappears down through the underlying Inferior Oolite to the relatively pervious Upper Lias Clay. To the west of this hole a depression running westwards may be traced which rapidly develops into a valley, which Richardson suggests pursues an underground course indicated by the dry valley described above.

Richardson (1933, p.28) noted at Oak Well, near Pool Keynes, that water sinks into the Cornbrash, resting on Forest Marble, near to the old railway. In the southern Cotswolds, Ackerman and Cave (1967) and Murray and Hawkins (1973) have noted swallow holes, developed in the Inferior Oolite of the Francombe Wood Valley.

A number of karstic depressions have been noted by Goudie (1967) which are as follows: a shallow depression, possibly of solutional origin at Jackament's Bottom (SO 965 975) near Kemble; a small swallowhole near a spring at Combend (SO 861 116), where a small spring disappears abruptly, and a hollow on Cleeve Hill, with a diameter of 20m, near to the Isaac's Cave system.

Goudie *et al.* (1980) mention that a number of caves occur in the Inferior Oolite, notably at Isaac's Cave, Cleeve Hill. On the scarp crest above Stanway, groundwater solution working along a small fracture has created a sizeable underground water channel (Beckinsale and Beckinsale, 1976). They also noted that such underground features included large scale rotting or weathering of limestone by waters rich in carbon dioxide. Two large pot holes, near Blockley, Gloucestershire, have been discussed by Beckinsale (1982b), though one has been filled with refuse and is barely visible. They occur in the escarpment summit of the north Cotswolds and have developed in the Inferior Oolite. The unfilled hole has been described in detail, and is in excess of 16m to its present rubbly infill. The depth of the infill is unknown.

Water disappears underground in the stream beds of the Cotswold rivers as has been observed in the R. Coln by Dury (1955). He observed a swallowhole about 750m above Withington, on the right hand bank of the stream, which had developed in the Inferior Oolite. Bravender (1867, 1868) considers that in the R. Churn, around half of the water lost after evaporation sinks into the Inferior Oolite, and he suggested that losses in the R. Coln valley may be greater than that for the Churn.

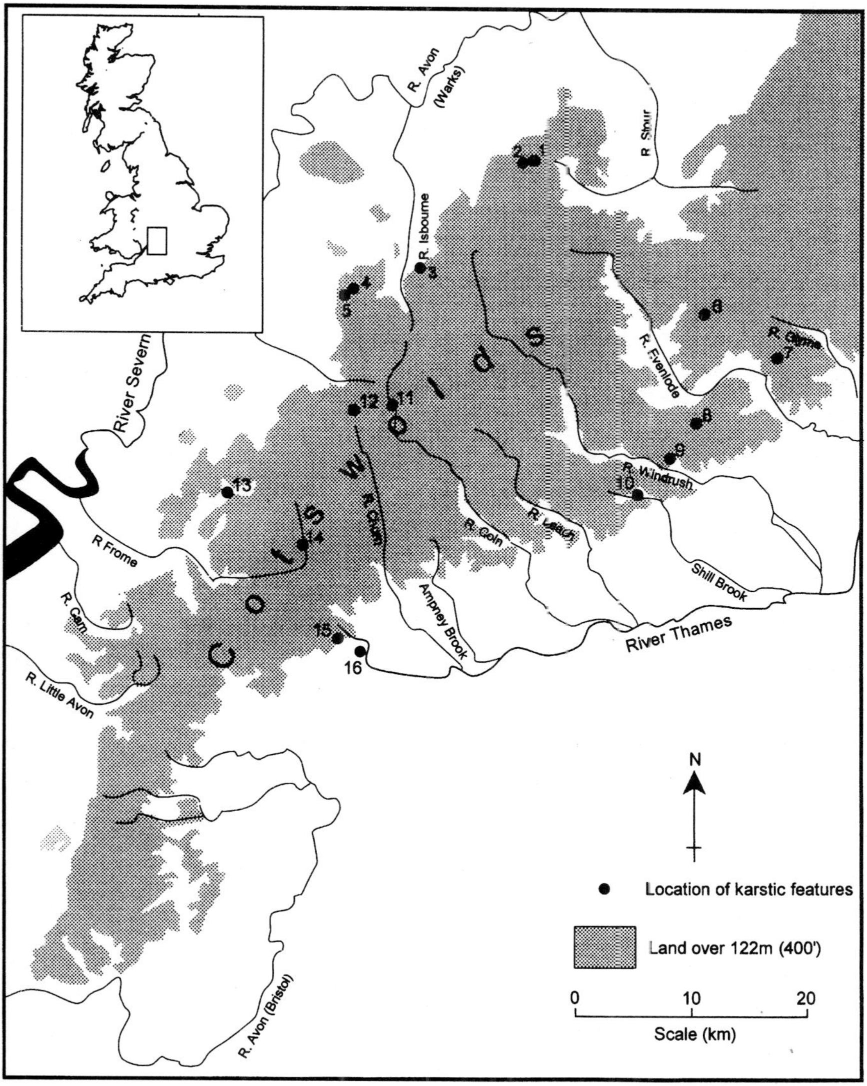

Figure 25 Location of karstic phenomena in the Cotswolds

Figure 26 shows the losses from the main Cotswold streams (after Beckinsale, 1970). Richardson *et al.* (1946) have also decribed 'tub-holes' 1km east of Asthall church from which water, presumed to be from the Windrush, emerge into Doctor's Ditch. Many of the dip slope streams of the Cotswold cuesta have discontinuous drainage with underground courses e.g. the R. Leach, Shill Brook and Ampney Brook (Goudie *et al.*, 1980).

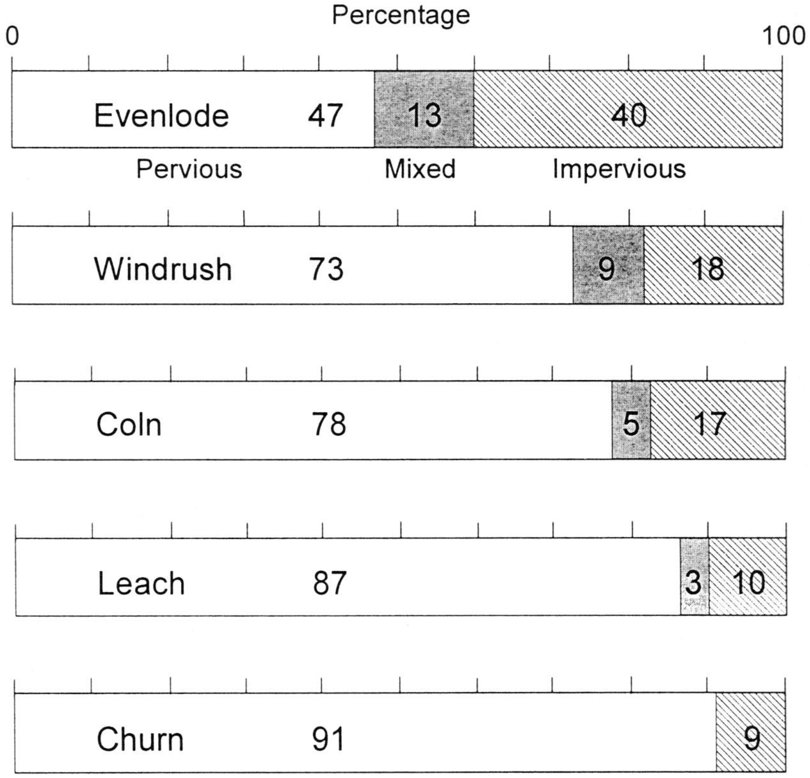

Figure 26 Present areal extent of limestone and clay outcrops in the drainage basins of some Cotswold rivers (after Beckinsale, 1970, fig. 3)

Moreover, some of the streams, even if they do not possess dry stretches, appear to suffer discharge diminution as they pass over the Inferior Oolite. Some early data on the Churn in 1859 were provided in Phillips (1871, p. 31). As it left its Lias bed, discharge declined from 320 cubic feet in a minute (at 5½ miles from its source) to 10 cubic feet in a minute (at 14½ miles from its source).

More recent data supporting this have been provided for the River Leach by M. G. Hart. He studied the hydrology of the River Leach in some detail during the period August to December 1972 and two important and contrasting events occurred at this time. In early November the river reached an unusually low stage, and one month later it flooded (Hart, 1976).

In the Leach Valley the effect of the surface outcrop of Inferior and Great Oolites on the stream is seen in its most extreme form, for at times of low flow the stream becomes intermittent. In November, 1972, for example, the channel was dry for about 14km from just below Larkethill Wood to just above Eastleach village. The dry reach so formed coincided approximately with the surface outcrop of the Great Oolite. At its lower end the dry reach was terminated by some muddy ground with occasional stagnant pools of deep water, and the stream was flowing at the surface again in Eastleach village.

In December heavy rains produced floods along most of the river, including the reach developed in the Great Oolite that had been dry one month earlier. At first the flood wave passed gradually downstream. Measurements indicated that the flood wave reached Lechlade two weeks after the the very heavy rainfall of that initiated the flood, and took six days to clear the lower 15km or so of the stream alone. It is usual, however, for flood waves to pass only slowly downstream on permeable lithologies.

The measurements of discharge showed that during the flood there was a gradual increase in discharge downstream, even along the reach that was dry at low flow. In places the discharge was lower than that to be expected in view of the additional increment of drainage area acquired downstream. In other words, when the Leach is in flood on the Great Oolite part of the catchment the progressive loss of water underground is delayed. Thus, the Cotswolds exhibit drainage systems and landforms characteristic of both karst and non-karst areas.

Goudie *et al.* (1980) account for the paucity of karstic forms in the Cotswolds by suggesting that the rocks themselves lack sufficient mechanical strength to support large-scale solutional features. This they suggest as being one of the most important factors in the lack of underground or collapse forms in comparison with the classic Carboniferous karst terrains. Studies in the water chemistry of springs and streams (Goudie, 1967; Goudie *et al.*, 1980, Paterson, 1970) show that the lack of karstic phenomena cannot be explained in terms of slow rates of solution (see below), for rates in the Cotswolds are of the same order as those in more 'classic' karstic terrains. Another important reason for the lack of karstic features is that the area has virtually no cover of non-calcareous rocks which can generate surface runoff and acidulated water in the same way that, for example, Old Red Sandstone does in the Mendips.

4.8 Spring water chemistry and rates of limestone solution

Springs, on account of the alternation of permeable limestones and impermeable clays, are widepread throughout the Cotswolds and the study of their water chemistry enables an assessment of the importance of solutional activity. If one calculates the amount of calcium and magnesium carbonate dissolved in stream water, removes the amount of carbonate coming into the area in precipitation, and estimates the quantity of discharge of the spring or stream catchment, one can calculate the amount of limestone being dissolved and removed per unit area per year.

As Tables 8 and 9 show, Cotswold springs have a calcium carbonate content that ranges between 123 and 413 parts per million (ppm). Other components are of markedly lower importance. Using available data on spring chemistry and stream discharge, rates of solutional denudation in the area would appear to range between 35 to 66 $m^3/km^2/yr$ as shown in Table 10 (Goudie, 1967; Paterson, 1970; Smith, 1965). These rates are comparable to those determined for other limestone areas in Britain and Ireland (Table 11).

Table 8 Calcium carbonate values for Cotswold springs (1965 and 1966) (Goudie, 1967)

		1965					1966					
Location	G.R.	Aug	Sep	Oct	Nov	Dec	Jan	Feb	Mar	Apr	July	Dec
Chedworth Manor	052121	305	257	264	220	320	321	-	-	-	-	-
Chedworth Villa	053135	220	-	-	-	-	-	-	-	-	-	-
Cleeve Nutterswood	982255	-	-	-	-	-	-	-	-	-	148	148
Cleeve Postlip	996262	128	129	123	123	123	-	-	-	-	-	-
Cleeve Rising Sun	984269	-	-	-	-	-	-	-	-	-	190	195
Cleeve Table	983249	-	-	-	-	-	-	-	-	-	337	353
Coberley	962160	-	-	-	-	-	-	-	-	-	217	220
Combend	984118	-	274	-	-	-	-	-	-	-	-	-
Compton Abdale I	061165	249	142	238	236	264	265	-	258	264	251	263
Compton Abdale II	060165	-	-	-	-	-	-	-	305	-	287	303
Compton Abdae III	063164	-	-	-	-	-	-	-	-	-	250	263
Upper Cowley	961142	-	-	-	-	-	-	-	-	-	276	339
Lower Cowley	957138	288	-	-	-	-	-	277	322	-	274	-
Crippets	938182	-	-	-	-	-	-	-	-	-	287	320
Guiting Power	095250	198	197	190	150	197	219	-	-	-	190	192
Hampnett	099159	-	-	-	-	-	-	-	-	-	287	320
Hilcot	997168	-	-	-	-	-	-	-	-	-	175	169
Kineton Ford	087260	230	258	285	229	275	-	-	-	-	-	-
Lower Swell	174256	210	209	209	170	229	203	-	-	-	209	223
Salperton	075209	-	-	-	-	-	-	-	-	-	240	285
Sandy Well	009 203	-	-	-	-	-	-	-	-	-	293	285
Seven Springs	967169	211	215	209	210	210	208	208	210	208	215	210
Shipton Oliffe	041188	-	-	-	-	-	-	-	-	-	284	285
Spring Bottom	966208	413	388	370	370	402	-	355	401	401	393	400
Ullenwood I	933175	-	-	-	-	-	-	-	-	-	288	303
Ullenwood II	933173	-	-	-	-	-	-	-	-	-	293	231
Ullenwood III	932171	-	-	-	-	-	-	-	-	-	231	245
Whittington	010211	-	-	-	-	-	-	-	-	-	398	338

Note: values obtained by E.D.T.A. titration

Table 9 Water chemistry data for different lithologies in the Cotswold region

	Hardness (ppm)					
Area	Total	Calcium	Magnesium	No. of. Samples	Ca/Mg ratio	Source
North Oxfordshire Heights (Jurassic)	344	334	10	392	33.40	Paterson (1970)
North Cotswolds (Jurassic)	-	257	-	109	-	Goudie (1967)
South Cotswolds (Jurassic)	-	260-370	-	-	-	Smith (1965)

Table 10 Rates of solutional denudation in the Cotswolds

Area	Rate ($m^3/km^2/yr$)	Source
Jurassic limestone of the North Cotswolds	35	Goudie (1967)
North Oxfordshire Heights (Jurassic)	60-66	Paterson (1970)
Thames Basin at Eynsham	60	Douglas (1964)

Table 11 Rates of solutional denudation in Britain exclusive of the Cotswolds

Area	Rate ($m^3/km^2/yr$)	Source
Peak district	75-83	Pitty (1968)
Craven	40	Sweeting (1965)
Fergus and Shannon	51-53	Williams (1963)
East Anglia	25	Perrin (1955)
Mellte, S. Wales	16	Groom and Williams (1965)
Lismore, Scotland	150	Corbel (1959)
Lismore, Scotland	100	Smith (1965)
Mendips	40	Corbel (1959)
Berkshire Downs	67	Paterson (1970)
Corallian Scarp (Oxon)	74	Paterson (1970)

Chapter 5

5.1. Mass movement phenomena and superficial structures

The broad spectrum of gravity-dominated slope movements are collectively known as movements.

The Mesozoic strata of the Cotswolds have in many places undergone superficial disturbance which is thought to have occured mainly during the freezing and thawing of deeply frozen strata during the Pleistocene (Cave, 1977). The steep slopes in the upland valleys and on the escarpments have in the past been relatively unstable, there being a range of superficial structures. Such structures include landslip, slip troughs, camber, dip and fault structures, gulls and valley bulges, and are well shown within the major river valleys and on the scarp and associated outliers of the Cotswolds.

5.2 Landslips

Jones and Lee (1994) have stated that

> "The Cotswolds come out as one of the main concentrations of known landslides in England and Wales."

The escarpment slopes and outliers represent the major concentration of landslips in the Cotswolds. Table 12 shows the frequency, density and type of landslides in Britain on strata which are to be found in the Cotswolds.

Among the earliest workers on the landslips was Murchison (1844, p. 59), who drew attention to a recent slip of a large knoll of Upper Lias occasioned by heavy rain at the Hewletts near Cheltenham. However, Murchison believed that some of the landslips were of 'high antiquity' and had a marine origin and that they were:

> "...the remains of former undercliffs which originated during that period (immediately anterior to the last desiccation of our island), when a great bay or channel of the sea occupied the broad valley of the Severn, and when the waves beat against the Cotteswold cliffs on the east, and the Malverns and May Hill on the west."

He believed, in short, that they were of comparable origin to the undercliff currently forming on the Isle of Wight. Murchison was writing at a time when such marine ideas were widely accepted, but there is no evidence to support this conclusion.

Table 12 The frequency, density and type of landslide characteristics of Jurassic strata typical of the Cotswolds (after Jones and Lee, 1994)

Stratigraphic Description *(Source: BGS 1:625,000 scale maps)*	**Number of recorded landslides**	**Total outcrop area (km²)**	**Density (total area) per 100 km²**	**Actual outcrop area (km²)**	**Density (actual area) per 100 km²**	**Unspecified landslides**	**Rockfalls**	**Topples**	**Sagging failures**	**Single rotational**	**Multiple rotational**	**Successive rotational**	**Compound**	**Translational**	**Flows**	**Complex**	**Cambered/foundered**
Cornbrash	8	688	1.2	634	1.3	4	-	-	-	1	-	1	-	-	-	-	2
Great & Inferior Oolite including Great Estuarine Series	68	1360	5.0	932	7.3	45	2	-	-	1	-	2	-	1	1	12	4
Great Oolite	176	2139	8.2	1767	10.0	128	2	-	-	1	15	-	-	3	11	9	7
Inferior Oolite	397	1819	21.8	1598	24.8	111	-	-	-	42	17	95	4	14	7	23	84
Undifferentiated Lower Jurassic	20	95	21.1	86	23.3	15	1	-	-	1	-	-	-	-	-	1	2
Upper Lias	662	1920	34.5	1591	41.6	316	-	1	-	50	14	96	10	14	50	29	82
Middle Lias	344	1417	24.3	1228	28.0	272	1	-	-	9	3	11	-	7	7	21	13
Lower Lias	310	5573	5.6	3235	9.6	191	9	-	-	14	15	2	4	15	17	39	4

Instability in the Cotswolds is produced through the combination of erosional lowering of the adjacent clay vale, thereby leading to slope steepening, and water emerging on the scarp face from the contact between the permeable caprock and impermeable lower horizons. The gentler the dip of the strata the greater is the proportion of groundwater that will emerge at the scarp face and therefore the more instability that will be generated.

Within the Cotswolds the two formations usually responsible for promoting landslips are the Lias, particularly the Upper Lias, and the Fuller's Earth (Lower Fuller's Earth Clay). Other minor occurrences of landslip have been recorded on the slopes of Forest Marble.

Multiple rotational forms are widespread (Jones and Lee, 1994). Often they attain considerable size as can be found on Bredon Hill, (Figure 27) (Morris, 1974; Gerrard and Morris, 1980; Whittaker and Williams, 1973), the Cotswold escarpment, especially at Cleeve Hill and Dovers Hill (Butler, 1983) and in the Bath area at Beacon Hill, Bailbrook and North Stoke (Kellaway and Taylor, 1968; Hawkins and Privett, 1979; Forster *et al.*, 1985).

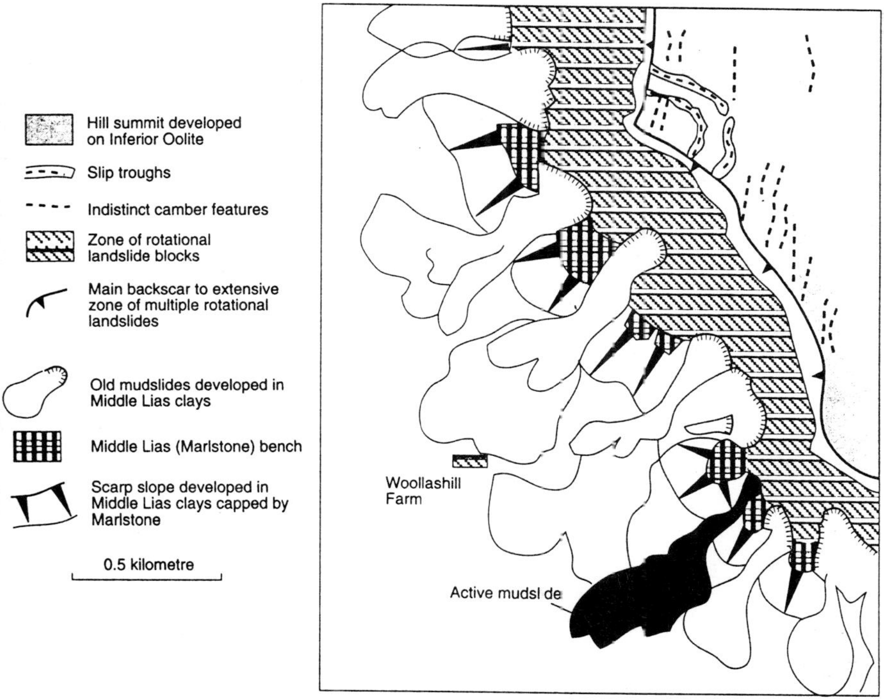

Figure 27 The pattern of landsliding on the western flank of Bredon Hill (after Morris, 1974)

Butler (1983) has noted that along the Cotswold escarpment successive rotational slides are found on the lower slopes below large-scale rotational failures. Successive rotational slides differ from multiple rotational slides in that in the former the slides are usually small, shallow failures confined to the weathered mantle and are characteristic of degrading slopes developed in thick fissured clay (Jones and Lee, 1994). The initial multiple failures of the escarpment degrade through a sequence of progessively shallower, part-succesive rotational failures and part translational failures, until a stable angle is achieved, as demonstrated in Figure 28 (Butler, 1983).

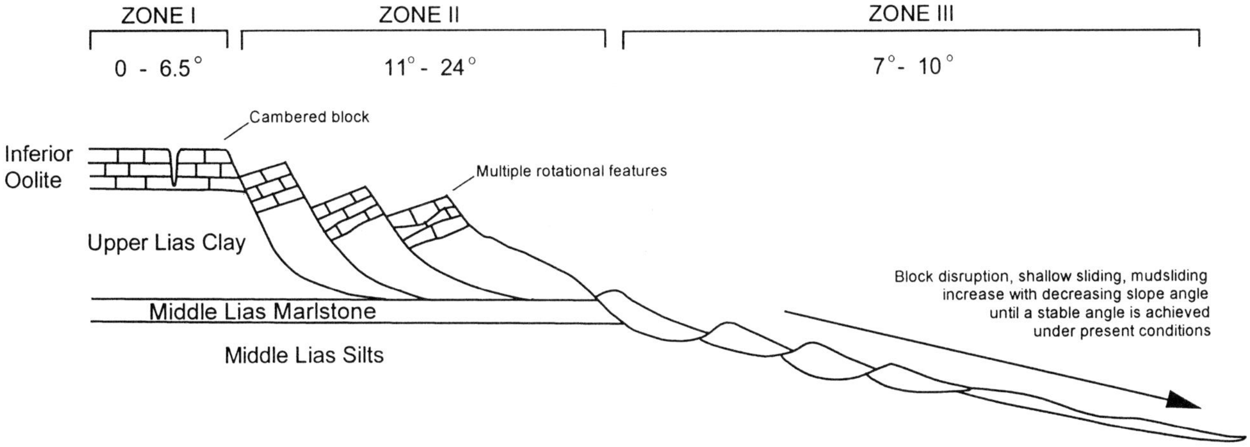

Figure 28 The degradation of valley slopes in the Cotswolds, showing the relationships between different landslide types (after Butler, 1983)

At Churchdown Hill, Gloucestershire, elongate mudslides have developed in Middle Lias Clays. They are long (800-2000m), narrow features found on long slopes which are steeper than the limiting angle for movement. Figure 29 shows the slide tracts and headscarps at Churchdown Hill (Hutchinson, 1991). Hutchinson (1967) has suggested that such slopes, which have developed in stiff clays, owe their present form to periglacial slope processes and have undergone little modification under temperate conditions during the Holocene.

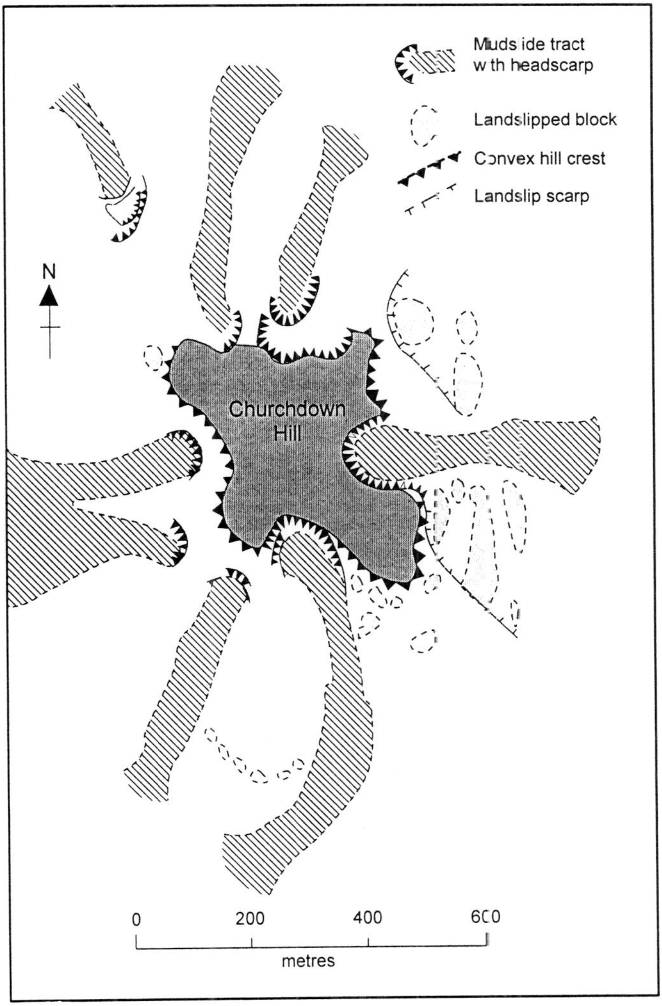

Figure 29 Churchdown Hill, relict periglacial mudslides developed over Lias Clay (after Hutchinson, 1991)

Throughout the Cotswolds there is evidence of shallow landslide activity since the Middle Ages. Goudie (1983) suggested that this was associated with the expansion of 'ridge-and-furrow' cultivation during the 'little optimum' of AD 750-1300. Evidence for this can be seen at Ebrington Hill and Leckhampton Hill where several areas have been disrupted by landslides. Butler (1983) mapped this at Ebrington Hill as shown in Figure 30.

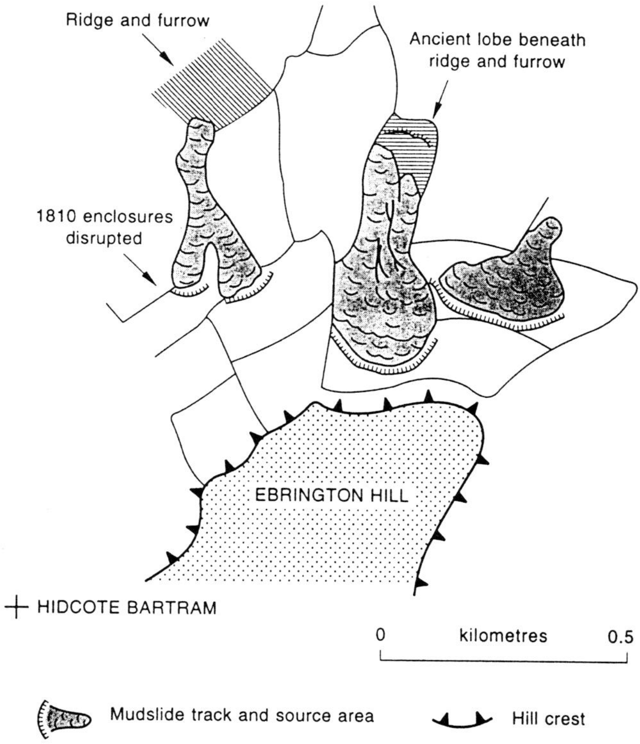

Figure 30 The pattern of post-enclosure (c. 1810) shallow activity on the northern slopes of Ebrington Hill, Gloucestershire (after Butler, 1983)

Although the Cotswold escarpment is as a whole known to be mantled by landslides (Butler, 1983). There are regional differences in their frequency. To the south of Stroud the escarpment is relatively free but to the north the area is heavily affected. This is due to a change in the geological conditions of the rocks which form the escarpment (Figure 31). In general, landsliding occurs on the Lower Fuller's Earth Clay, a 10m thick calcareous, silty, overconsolidated mudstone with poorly developed, very thin bedding but with a close network of stress release fissures. The physical properties of this material makes it the least stable slope-forming material in the area (Hawkins *et al.*, 1988). To the south of Stroud the Fuller's Earth Clay is sandwiched between the stable Inferior Oolite and the overlying Great Oolite cap-rock, so that failures only occur where the cap-rock is thin or dislocated.

However, to the north of Stroud there are marked geological changes in the rock types present in the scarp. The Great Oolite and Fuller's Earth Clay disappear whilst the Inferior Oolite thickens northwards and becomes the main scarp former. Beneath the Inferior Oolite, lies Upper Lias Clay which forms a substantial thickness on the scarp face beneath the Inferior Oolite, and which promotes landsliding. Butler (1983) has noted that in the areas where Upper Lias outcrops in the scarp 51% is mantled by landslides.

5.3 Cambering, valley bulges, gulls

Cambering, valley bulges and associated features were first described in the late 18th and early 19th centuries, mainly from areas of northern England e.g. Miller (1887), Watts, (1905) and Lapworth (1911). Hollingworth *et al.* (1944) wrote their classic paper on cambering and similar features from the Jurassic rocks of Northamptonshire. Cambered strata are perhaps the most important superficial structures to be found on the Jurassic rocks of the Cotswolds. Within the process of cambering a number of associated features are produced, which shall be described below.

Cambering is the downlope bending of a rigid surface of rock on the flanks of valleys (valley cambering) or on the crests of escarpments. It is best developed in those situations where valleys have been cut into sub-horizontal strata consisting of a rigid cap-rock, a layer of clay and an underlying rigid sub-stratum.

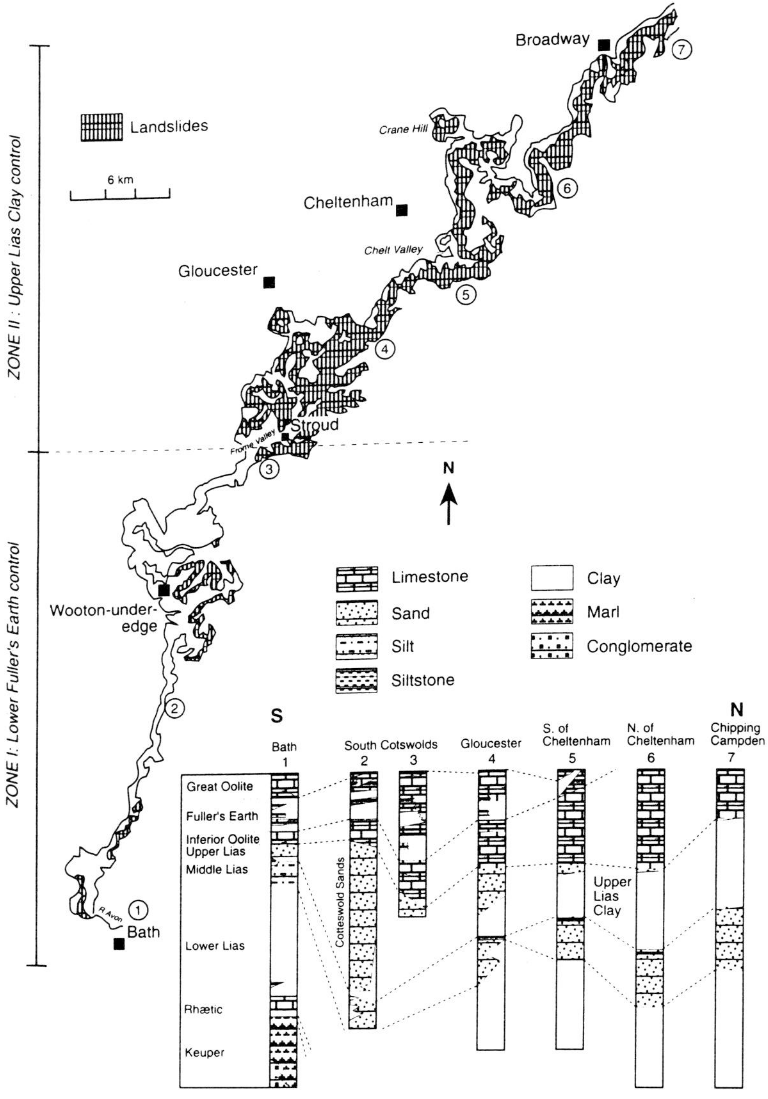

Figure 31 The importance of lithology in controlling landsliding along the Cotswold escarpment (after Butler, 1983)

Cambering is topographically important in that otherwise angular hilltops become rounded, and in that it tends to give misleading impressions of an excessive thickness of the surface capping, to ridges and small plateau areas. Figure 32 (after Parks, 1991) shows a general model of cambering and valley bulging under periglacial conditions.

Within the northern Cotswolds cambering is most apparent in the areas of Middle Lias where the hard Marlstone Rock Bed is thick and underlain by clays (Horton *et al.* 1987). Perhaps the most spectacular example of cambered strata is displayed at Bredon Hill, where the base of the Inferior Oolite drops over 30m down the Upper Lias slope of less than 140m (Williams and Whittaker, 1974). At Dover's Hill the Inferior Oolite drops 21m over a distance of 450m. Intensive cambering is also displayed at Ebrington Hill. In the Swainswick Valley, near Bath, Chandler *et al.* (1976) described clays which were brecciated to a depth of 40m within cambered strata (Figure 33). The influence of cambering on local land-use and the development of Bath has been studied and described by Kellaway and Taylor (1968) and Hawkins and Privett (1981).

Camber slopes are also found in the Windrush Valley. Worssam and Bisson (1963) have described these features between Windrush and Burford, which mainly occur where the Inferior Oolite Series (especially the Clypeus Grit) overlie the Middle Lias Clays which floor the valley of the river. The cambering is only really apparent from exposures within quarry sections, notably at Pinchpool Farm Quarry at Windrush, Kitt's Quarry, Upton and Lee's Quarry, Taynton. Two outliers of Clypeus Grit, ¾ km east of Great Barrington are suggested as being the result of the dissection of a camber slope (Worssam and Bisson 1963, p.112).

Arkell (1947a, p.150) noted extensive cambering in the valley of the R. Evenlode, especially between Chadlington and Fawler in which both sides of the plateau of the Clypeus Grit, Chipping Norton Limestone and Taynton Stone (80-90 feet thick) slope gently towards the river. Springs along the interface between the limestone and Upper Lias Clay have been extremely active and have cut back a number of steep, narrow-sided valleys, which has destabilised the slopes.

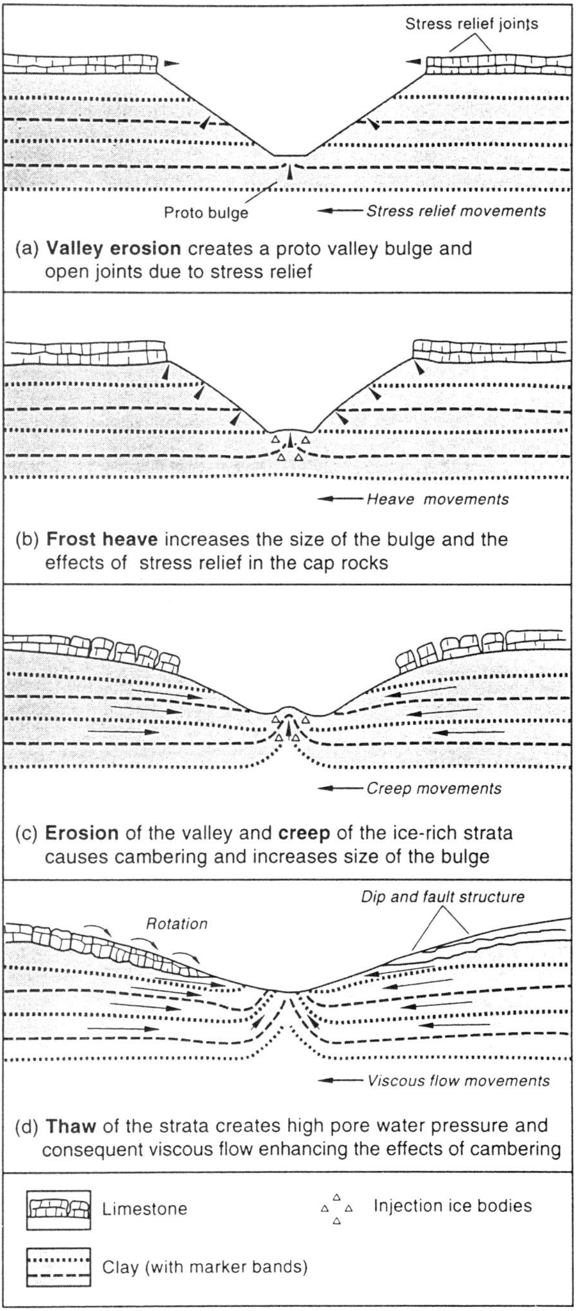

Figure 32 The development of cambered structures and valley bulges (after Parks, 1991)

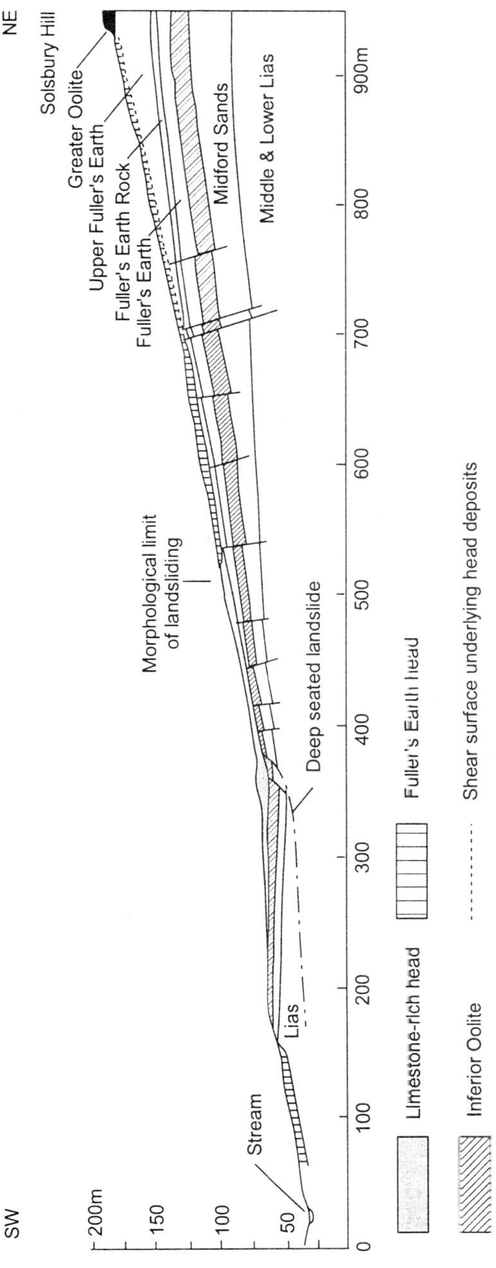

Figure 33 Section through Swainswick Valley, near Bath (after Chandler et al., 1976)

Gulls are essentially widened joints formed by the extensional movements of rocks affected by cambering. The tension created by the downward bending causes pre-existing major joints to open, thereby sub-dividing the cambered layer into individual blocks. It is common for gulls to be aligned parallel to the principal joint orientation. Gulls are normally filled with overlying superficial deposits which collapse into the voids as they open. Infilled gulls associated with cambering occur to the west of Stow-on-the-Wold (Briggs and Courteney, 1972), at Bredon Hill (Whittaker, 1972), and at Crickley Hill. Deep gulls have also been described at Radstock, near Bath (Hawkins and Privett, 1981).

Large gulls occuring in the Great Oolite, filled with subsided Oxford or Kellaway Clays and Northern Drift were noted at Breakspear's Quarry, East End, near North Leigh, by Arkell (1947a). These gulls have formed at the breach of large detached slices in the valley side.

Lowering of the strata on the uphill side often occurs when there has been vertical displacement across a gull or series of gulls along a slope profile. The result of this will be to produce a down hill dip greater than that of the regional dip. Thus dip and fault structures arise. Dip and fault structures have been noted in the valleys of the Rivers Windrush (Worssam and Bisson, 1963) and Evenlode (Arkell, 1947a).

Valley bulges consist of strata that have bulged up in the base of a valley as a result of erosive processes (Goudie, 1994). Valley bulges usually take the form of broad anticlinal deformations of the argillaceous strata underlying the valley floor, with the fold axis running parallel to the valley axis. Often the tops of these bulges have been eroded away and they are normally part obscured by alluvium. The beds within a bulge structure are highly sheared and softened, with the base of the deformed zone marked by one or more zones of décollement. Hawkins and Kellaway (1971) and Kellaway *et al.* (1971) suggest these structures are Pleistocene in age and were promoted by snowmelt and thaw giving high pore pressure conditions compared with today.

They are widespread in the sedimentary rocks of the English Midlands generally, but are especially well developed in the Cotswolds, where limestones and clays occur in close juxtaposition. Ackerman and Cave (1967) described examples from the valley of the River Frome near Stroud (Figure 34).

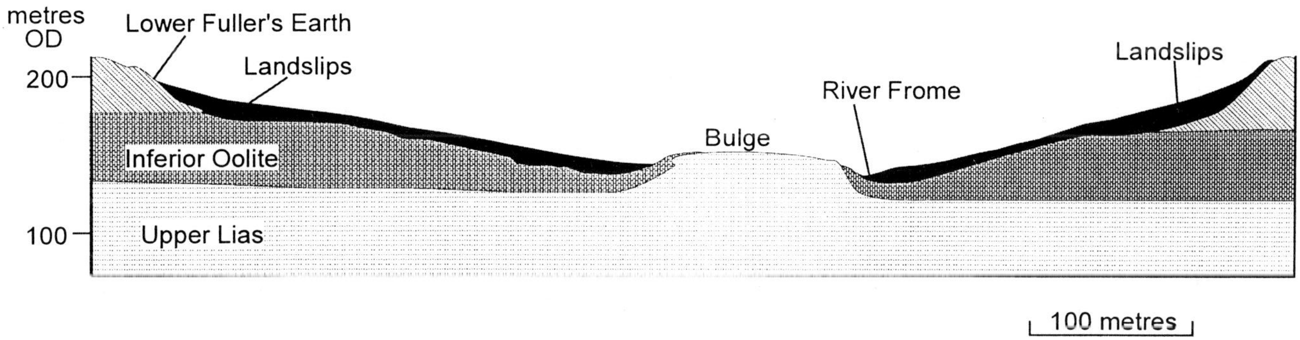

Figure 34 Valley bulge structures in the Frome Valley, near Stroud (after Ackerman and Cave, 1967)

The mechanism which they invoked was that during the Pleistocene severe erosion and valley incision occurred during permafrost conditions. As the rocks thawed at the end of the cold period, the Upper Lias clays, silts and sands were highly charged with water, became plastic, and under the weight of the more competent limestones above were extruded up through the weaker points of the recently developed valley floors. At Pinbury Park, the 'plugs' of sand and silt forced through the Inferior Oolite, which is 7m above the valley floor, were sufficient to block and then divert the course of the River Frome.

Arkell (1947a, p.151) has also described valley bulging in the Evenlode Valley in which

> "...squeezing and up-thrust structures in the valley floors were caused by differential loading of the plastic sub-stratum of Lias as a result of the surface relief. The requisite release of pressure in the narrow part of the Evenlode Valley was evidently afforded by the breach of the solid limestone cover."

Further upsteam at Chipping Norton, Lower Lias Clays have been brought to the surface by valley bulging (Arkell, 1947a, p.18).

5.4 Slip troughs and ridge and trough topography

Goudie and Hart (1975) have described a group of large relief forms produced as a result of cambering, namely slip troughs. At Cleeve Hill, near Cheltenham, cambering occurs in the Inferior Oolite overlying Upper Lias Clay (Figure 35).

Slip troughs are long (up to 500m) and deep (up to 14m) depressions, often with steep slopes (up to 20°) with broad bases (up to 70m wide), which at first sight resemble dry valleys. They do however, have unusual morphological characteristics which distinguish them from dry valleys. These characteristics recognised by Goudie and Hart(1975) are repeated *verbatim* below:

> (a) They run parallel to the main relief trends of the Cotswold escarpment and the major combes such as the Winchcombe embayment.
>
> (b) They appear in some cases to run across the normal dendritic dry valley systems such as the Postlip and Corndean valleys.
>
> (c) They have up and down long profiles or may sometimes be closed depressions like elongated dolines.
>
> (d) They have consistently asymmetric cross-profiles, the plateau-facing slopes being less steep than vale-facing slopes. At Lower Hill Farm (SO 992 240), for example, thirty

random measurements of maximum trough side slope angle showed that the plateau side of the trough had a mean declination of 20.7° (standard deviation 2.5) while the escarpment side was less steep (mean angle 15.7°, standard deviation 3.7). The difference is statistically highly significant.

(e) The troughs and depressions have no normal catchment areas.

(f) They have somewhat flat bottoms, which may contain a deep fill. Up to 5m of non-alluvial fill has been recorded by augering.

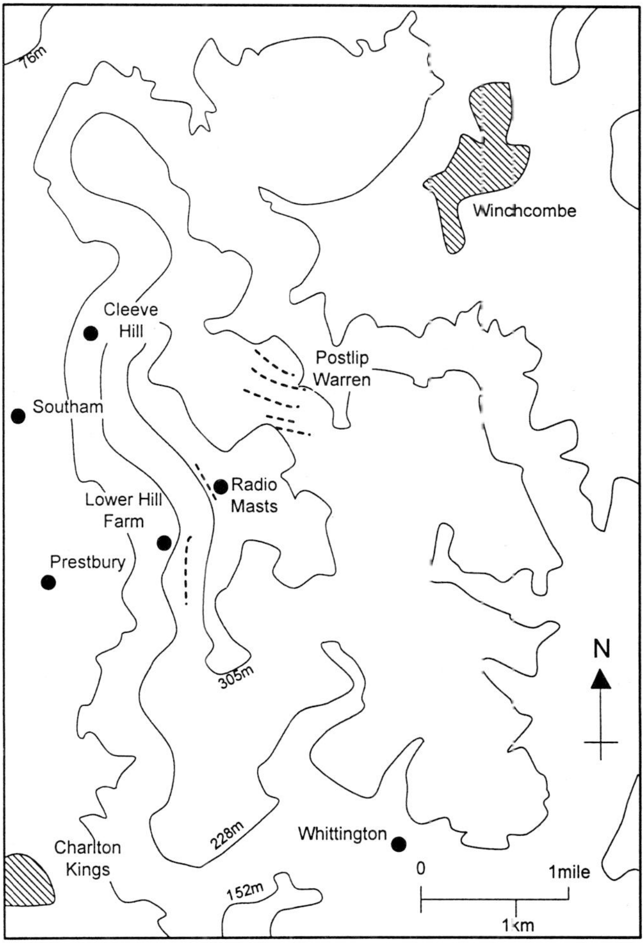

Figure 35 Location of major slip trough features on Cleeve Hill. The troughs are marked by dashed lines (after Goudie and Hart, 1976)

The main theory suggested by Goudie and Hart (1975) for these features was that the troughs are large scale slip troughs produced by the foundering of large masses of oolitic limestone over less competant Liassic clays and marlstones. Cambering is evident in many quarry sections on Cleeve Hill, and would explain

(a) Why the troughs develop parallel to the edges of the scarp, where oversteepening occurs.

(b) Why no catchment area is required.

(c) The asymmetry and steepness of the troughs.

(d) That the tilted blocks produced could provide the superficial material which fills the trough bases.

Morris (1974) has also mapped slip troughs on the western flank of Bredon Hill (see Figure 27).

In essence ridge and trough topography is the same as slip troughs and describes the ground surface form recognised on the interfluves between valleys. The most widely recognised examples in the British Isles were described in the Windrush Valley by Briggs and Courteney (1972) (Figure 36). At Upper Slaughter the troughs developed in the Inferior Oolite are 9m deep, up to 300m in length, and form a trellis like pattern of shallow depressions which hang above the main Windrush valley. Ridge and trough topography is essentially a well developed network of collapsed gulls which have been only partially infilled by debris, so leaving distinctive surface expression (Jones and Lee, 1994).

5.5 Solifluction spreads

The widespread solifluction spreads of the Cotswolds are a varied group of locally derived, unsorted or poorly sorted materials accumulated by a process of downslope sludging, the mobilisation being due to an excess of porewater. Within the broad context of solifluction spreads one can include Head and Fan Gravels. The process of solifluction appears to have been most active under periglacial conditions during the cold periods of the Pleistocene. For example Chandler *et al.* (1978) noticed clay-rich head deposits overlying organic deposits of Windermere Interstadial age (*ca.* 11-12 000 yr BP) from Horsescome Vale, near Bath.

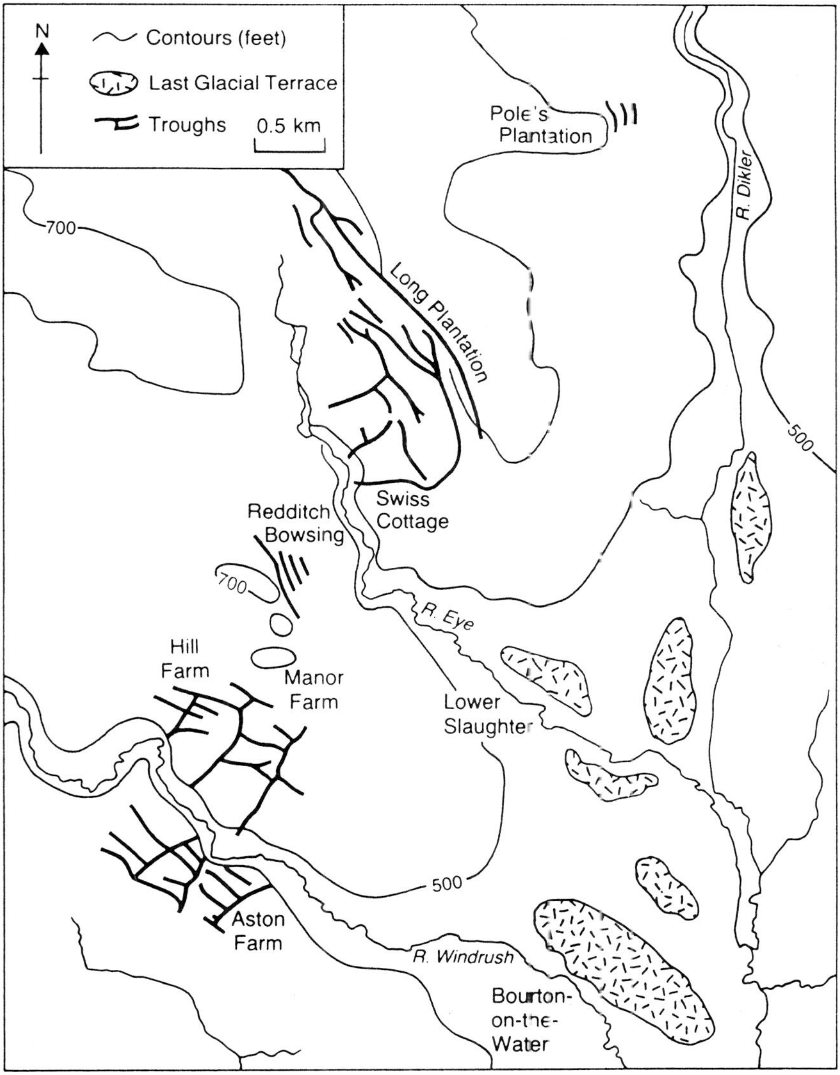

Figure 36 Ridge and trough features in the north Cotswolds (after Briggs and Courteney, 1972)

Periglacial slope deposits in lowland Britain are generally referred to as 'Head', a term introduced by De la Beche (1839) to describe the unsorted, unconsolidated 'rubble drift' of south west England (Ballantyne and Harris, 1994). Head deposits in general are formed primarily as a result of solifluction including gelifluction or frost creep, but may also have accumulated by shallow landsliding over permafrost (*sensu* Harris, 1987). The material incorporated in Head deposits is often derived from weathering of local bedrock, and the nature of this parent material often has a considerable influence on its sedimentological and geotechnical properties (Ballantyne and Harris, 1994). Often associated with many Head deposits is Brickearth or loess, which commonly consists of a mixture of silt-sized quartz grains which is aeolian in origin.

Opposite many of the larger embayments in the Cotswold escarpment, fans of angular Jurassic gravel spread out (Arkell, 1947a). The term Fan Gravel is used for solifluction detritus that is derived from the high ground of the northern Cotswold escarpment and outliers (e.g. Bredon Hill), and which has been re-sorted by fluvial action and deposited as fan shaped spreads on the gentler lower slopes. The deposits are mostly 1 to 2m in thickness, and may locally attain 3 to 4m. The deposits were mapped and described by Tomlinson (1941) between Chipping Campden in the northern Cotswolds and Stroud and Frampton in the southern Cotswolds. Some of the largest sheets occur around Cheltenham and Gloucester. At Cheltenham they contain much interstratified quartz sand of non-local origin, which is known as Cheltenham Sand (Arkell, 1947a) (see section 3.11).

Tomlinson (1941, p.385) described the fan gravels in detail in the area around Bredon Hill and Mickleton, following the edge of the low ground bordering the Cotswold escarpment (Figure 37). The spreads of gravel have been deposited as gently sloping fans, and are composed of Inferior Oolite limestone fragments up to 20cm across, angular to sub-angular with fragments of Marlstone Rock Bed, Lias and occasional Bunter and Flint pebbles. There is often a rough stratification and in places seams of quartzose sand with rounded gravels occur (Williams and Whittaker, 1974). The rounded sand is thought to be aeolian in origin. It is thought that the fan gravels formed under periglacial conditions as torrent deposits in which seasonal thaw or local snow and ice contributed to the run-off (Tomlinson, 1941; Williams and Whittaker, 1974).

The age of these deposits is thought to be Devensian. At Wingmoor Farm (SO943 270) working for gravel revealed a number of channels filled with organic silts and clay, which contained plant macrofossils and molluscs (Briggs, 1984). They have yielded arctic tundra faunas including mammoth, woolly rhinoceros, musk-ox, reindeer and lemming (Green, 1992). Pollen evidence suggested an open vegetation probably interspersed with areas of disturbed ground (Hunt, 1984).

In the Carrant Brook, radiocarbon dates of 27 000 to 28 000 years BP confirm a mid-Devensian age (Briggs *et al.*, 1975).

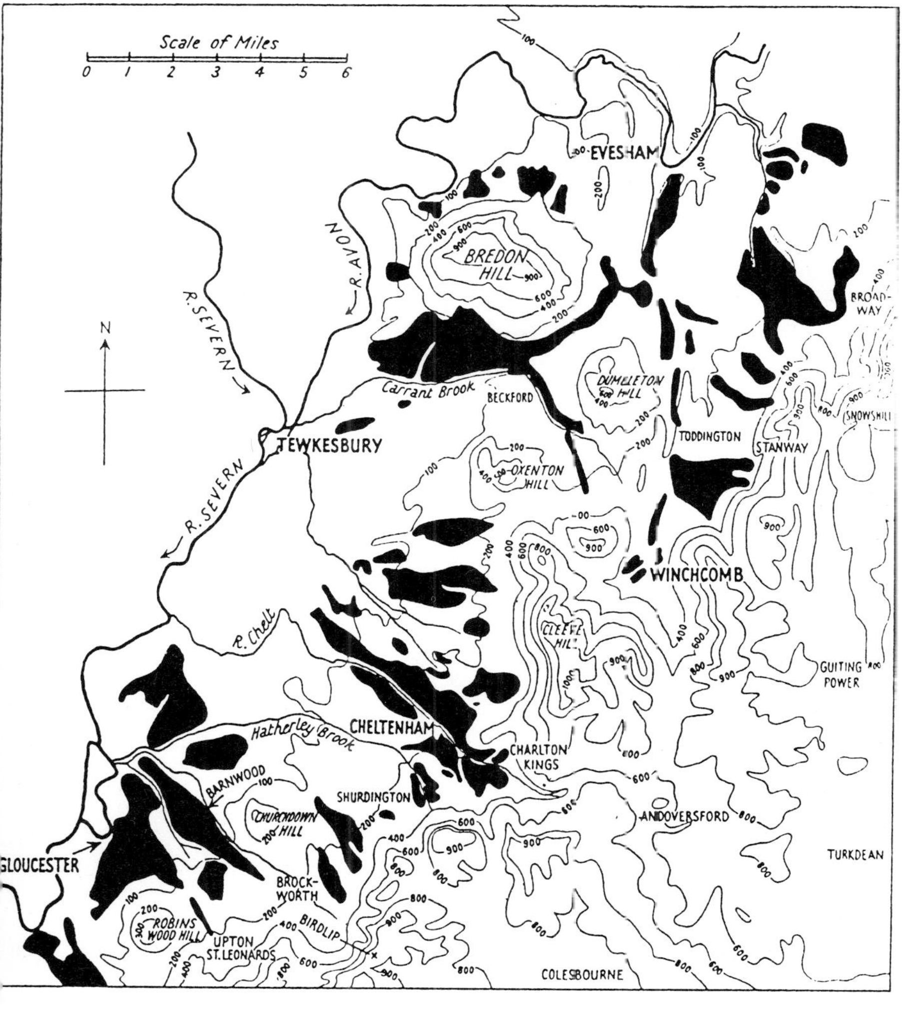

Figure 37 Map of the Fan Gravels at the foot of the Cotswold escarpment (after Tomlinson, 1941)

At Aston Magna a thick fan gravel overlies the Chalky Boulder Clay in the Moreton Gap (Arkell, 1947a). Fan gravels, up to 5m thick, have been described by Arkell (1947a) from Upper Milton gravel pit, Milton-under-Wychwood. The fan gravels near Milton are shown on the BGS

Witney sheet 236 at SP 261 182, SP 262 173 and a long tract along a tributary valley of the R. Evenlode running SW-NE from SP 258 175 to SP 271 181. Deposition dips of up to 12° have been measured (Arkell, 1947). Dury (1955) reports similar material from the R. Coln at Withington. This material is unbedded, with no indication of river deposition. On the left hand bank it is known to extend beneath a terrace-like feature *ca.* 0.75m thick. Dury suggests that the limestone debris sludged down, and was in part re-distributed by water. Richardson (1929) recorded what he termed 'downwash gravel' at Syreford, upstream from Withington. Again this material is unbedded with no signs of river deposition. In essence this is fan gravel material (*sensu stricto*). Dines (in Richardson, 1933), notes an abundance of such material in the neighbourhood of Coln St. Denis, together with exposures at Chedworth, where gravels are cemented in part and lie 4m above the present floodplain.

Large spreads of solifluction deposits occur north of Bredon Hill and the Cotswold escarpment, lower down the slopes from the Fan Gravels, and consist of brown loams and loamy clays, with scattered quartz grains and occasional Bunter and flint pebbles. In places they appear to overlie the Fan Gravels (Williams and Whittaker, 1974). Large accumulations of Head deposits border the lower slopes of the Edge Hill scarp in the north of the area (Edmonds *et al.*, 1965).

Chapter 6

6.1 The form of the escarpment and its outliers

The Cotwolds form a striking escarpment which runs from Bath to Meon Hill, a distance of 65km (Figure 38).

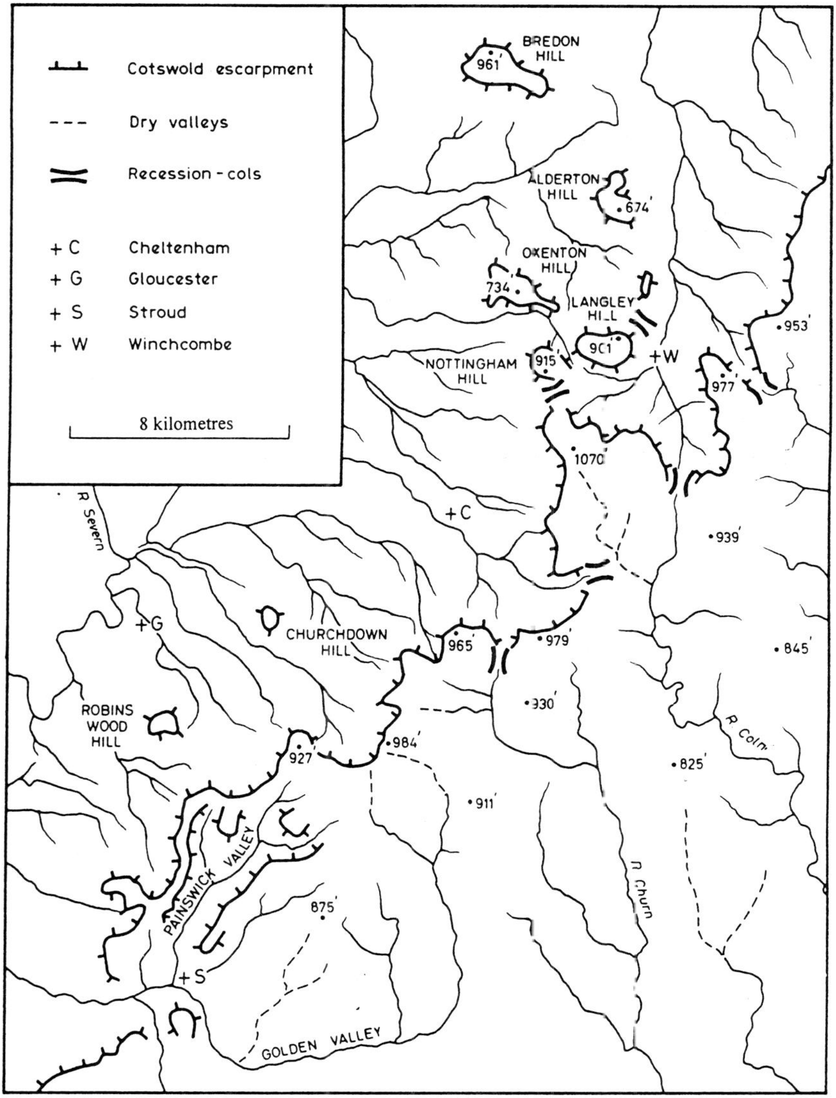

Figure 38 The Cotswold escarpment between Stroud and Winchcombe (after Small, 1970)

The hills tend to run south-west to north-east with their scarps facing north-west, and the orientation of the strata being to the south east. The cuesta is capped around Stroud and to the north-east by Inferior Oolite limestones and to the south by Great Oolite. The upper part of the scarp beneath the cap rock is formed by Upper Lias Sands (Cotteswold Sands) and Clay, and overlooks a structural bench formed by the uppermost division of the Middle Lias, the Marlstone Rock. This lower structural bench along the escarpment occurs between Wotton-under-Edge and Dursley and between Cheltenham and Chipping Campden. Figure 39 shows cross-sections across Leckhampton and Stinchcombe Hills (after Richardson, 1908). The lowest part of the scarp is developed in Middle Lias sands and clays, and Lower Lias Clay. The present position of the scarp is transitional and on this basis the cuesta is supposed to have extended further westwards and to be in the continuing process of retreat.

Jones and Lee (1994, p. 202) state that:

> "The prominence of the Cotswold escarpment, the outliers and the occurrence of landslides are all symptomatic of recent intense geomorphological sculpturing"

Early investigators of the escarpment thought that it had been moulded by marine forces (see, for example, Murchison, 1844). The most outspoken exponent of this viewpoint was Mackintosh (1869, p. 109 *et seq.*) who averred:

> "It is not more certain that the sea is in its present bed than its waves once washed the escarpment up to a considerable height."

He regarded, for instance, the Witcombe embayment as a 'great bay', the free faces of Crickley Hill as 'a range of oolitic cliffs', the debris at the escarpment base as a 'beach', the Devil's Chimney on the sides of Leckhampton Hill as 'a natural sea-coast needle', and the 'King and Queen rocks' on the side of Bredon Hill as 'sea-coast rocks which have escaped atmospheric denudation'. Such an opinion, not atypical for the mid nineteenth century, has seldom received serious support since that time.

Subsequent discussions by Davis (e.g. Davis 1899) made it clear that the Cotswold cuesta was a classic of the genre and to be compared with that of the Swabian Alb in southern Germany.

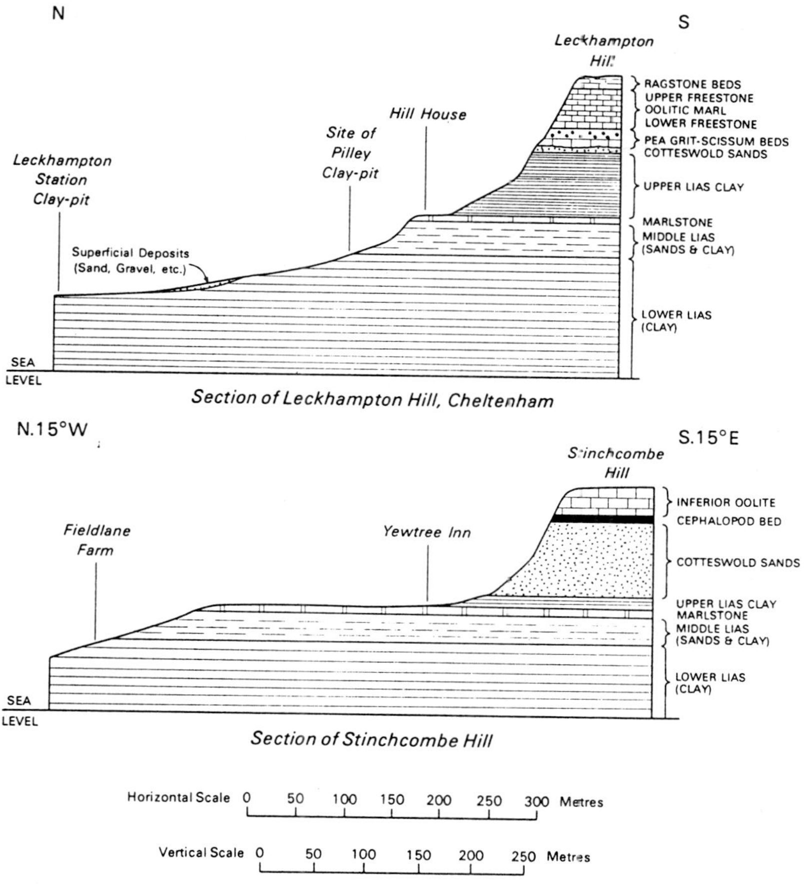

Figure 39 Sections across the Cotswold scarp (after Richardson, 1908)

The topography depends largely on the thickness and porosity of the strata and on the difference in resistance to erosion displayed by the clays and limestones. The Cotswold escarpment is fretted and in a number of places dissected by anti-dip streams, with notable examples around Stroud, where the Golden Valley penetrates several kilometres eastwards into the upland, and in its lower parts is joined by the strike-aligned Painswick and Nailsworth valleys. To the east of Cheltenham the Winchcombe valley, though broader and more mature than those at Stroud, forms a major re-entrant. Small (1970) has noted that the tributary streams at Winchcombe have, by headward erosion, almost succeeded in isolating the upland masses of Langley and Nottingham Hills. The steep, north-west facing escarpment is made irregular by a number of short, steep, scarp face valleys, such as that above Hailes (SP 058 285), Snowshill (SP 093 345) and at Kites Nest (SP 105 353) (Adlam, 1974), as well as the large Vale of Moreton.

Bowen *et al.* (1986) suggested that the evolution of the lower Severn valley, to the west of the present Jurassic escarpment of the Cotswolds, was dominated by three events:

1. erosion during the Anglian phase (480-428 000 years ago)
2. the formation of a huge pro-glacial meltwater lake between Bredon Hill and Leicestershire (see Figure 13 in chapter 3) This resulted in the diversion of the formerly eastward flowing drainage into the Severn (e.g. the Warwickshire Avon).
3. the diversion of the upper Severn drainage southwards through the Ironbridge Gorge

They suggest that the period of the Anglian glaciation included the erosion of the Severn Basin, and that it was then that the Cotswolds became prominent features. Thus prior to the Anglian, they argue the Cotswold escarpment would have been a minor feature. The post Anglian environmental history resulted in rapid landscape evolution by the enlarged and much invigorated River Severn. This would account for the the prominence of the Cotswold escarpment and provide conditions that favoured some of the most impressive landslide developments, as discussed in chapter 5.

The Cotswolds are remarkable in the sense that they have a number of completely detached outliers, which are in general rare elsewhere in Britain. Local examples include Meon Hill, Bredon Hill and Alderton Hill in the north of the Cotswolds, Oxenton Hill to the north of Cheltenham and Robins Wood and Churchdown Hills east of Gloucester. Figure 40 shows their positions and Table 13 summarises the main characteristics of the outliers. The outliers are based on those locations where hills have been separated from the main Cotswold mass down to the Lower Lias. Thus, many other sites such as Uley and Cam Long Down may at first appear to be outliers, they have not infact been separated down to the Lowere Lias. Cross sections across some of these outliers are shown in Figure 41. Hull (1857) suggested that the survival of all the Cotswold outliers was wholly dependent on structure, and the proximity of faults.

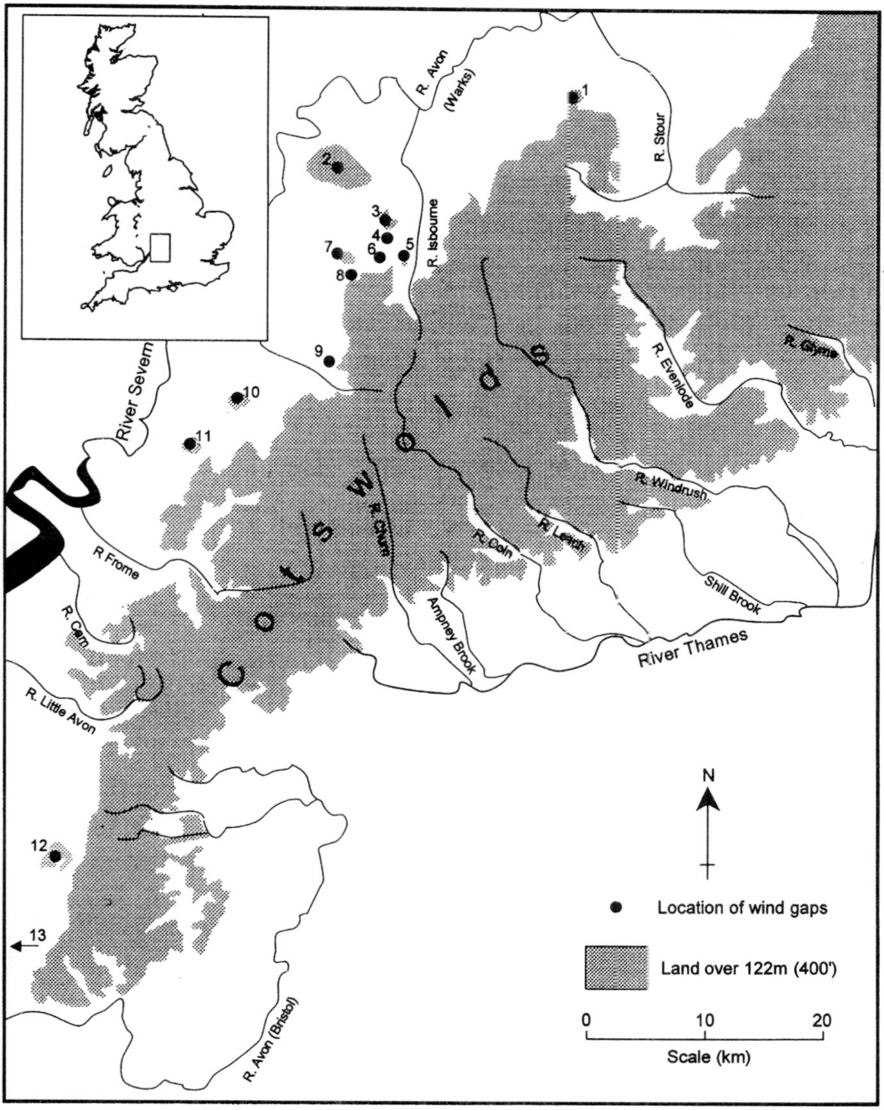

Figure 40 Location of the Cotswold outliers

Of the 13 outliers listed in Table 13, it can be seen that 4 are capped by the Inferior Oolite, 6 are capped by the Middle Lias, and 3 are capped by the Upper Lias. Of the outliers Bredon Hill is the largest (5.6 x 2.4 km), the highest (293m) of the outliers and the furthest removed from the main escarpment (7.75 km).

Table 13 Outliers of the Cotswolds

Outlier	Grid Ref.	Max. Alt. (m)	Geology	Dist. from Scarp (km)
Meon Hill	SP 178 453	194	Middle Lias	5.25
Bredon Hill	SO 970 395	293	Inferior Oolite Upper Lias Middle Lias	7.75
Dumbleton Hill	SP 010 355	168	Middle Lias	4.5
Alderton Hill	SP 010 345	206	Upper Lias Middle Lias	4.0
The Warren	SP 025 315	135	Middle Lias	2.0
Dixton Hill	SO 985 307	164	Middle Lias	1.2
Oxenton Hill/ Woolstone Hill	SO 972 314	222	Inferior Oolite Upper Lias Middle Lias	1.75
Crane Hill	SO 963 306	128	Inferior Oolite Upper Lias Middle Lias	1.5
Battledown	SO 970 220	140	Middle Lias	1.0
Churchdown	SO 850 189	137	Middle Lias	3.5
Robins Wood Hill	SO 841 151	198	Upper Lias Middle Lias	1.0
Bitton Hill	ST 680 705	80	Upper Lias	1.5
Dundry Hill	ST 553 667	233	Inferior Oolite Upper Lias Middle Lias	7.0

In this context, Small (1970,p.83) has suggested that outliers can develop only:

(i) where the angle of the dip is exceptionally low (that of the central Cotswolds is only 1°).
(ii) where scarp dissection is abnormally advanced (for the outliers are, in effect, the unconsumed remnants of interfluves between adjacent escarpment valley systems). Conversely they are virtually never present where the angle of dip of the cuesta forming rock exceeds 5°, or where the scarp-face undergoes more uniform or 'normal' retreat by mass wasting processes. Additional information relating to the theory of scarp recession is the nature of the wind-gaps breaching the Cotswold divide as described in chapter 4.

The unusual morphology of the Cotswold escarpment poses some difficult problems What is apparent is that certain individual valleys e.g. at Winchcombe and the Chelt valley, have been eroded back along lines of former dip-slope valleys, and that the agencies responsible have been able to take advantage of important recession-cols incised well below the general summit of the cuesta. A number of hills are almost completely detatched from the main escarpment e.g. Ebrington Hill, Langley Hill and Lansdown and in time will themselves become outliers. Scarp valleys, which in many other areas have been explained in terms of headward erosion by spring sapping, cannot explain for example features such as the Golden Valley at Stroud.

Figure 41 Sections across some of the Cotswold outliers

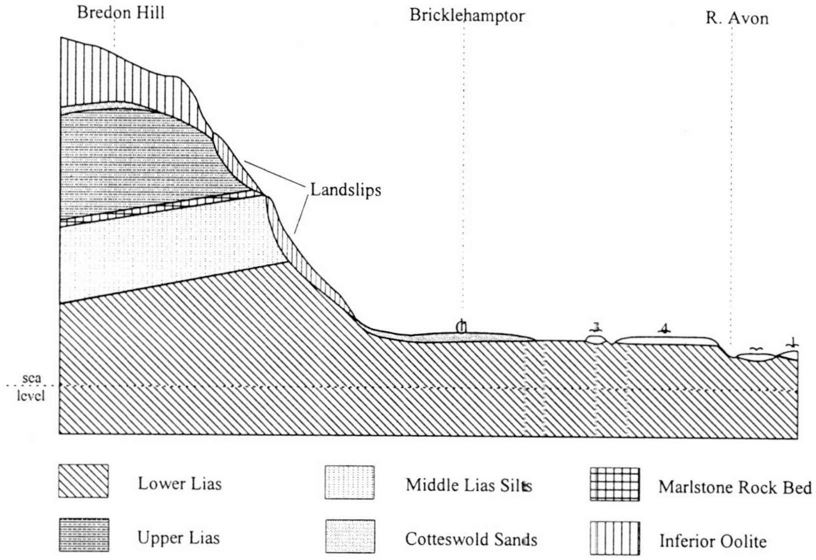

Section across Bredon Hill (after BGS)

Section across Oxenton Hill (after Hull, 1855)

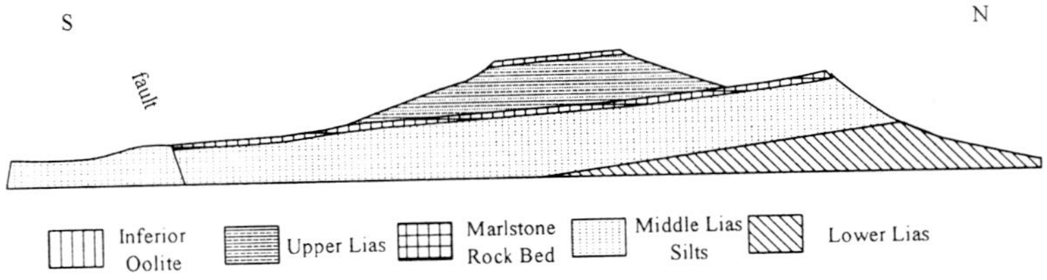

Section across Battledown (after Murchison, 1844)

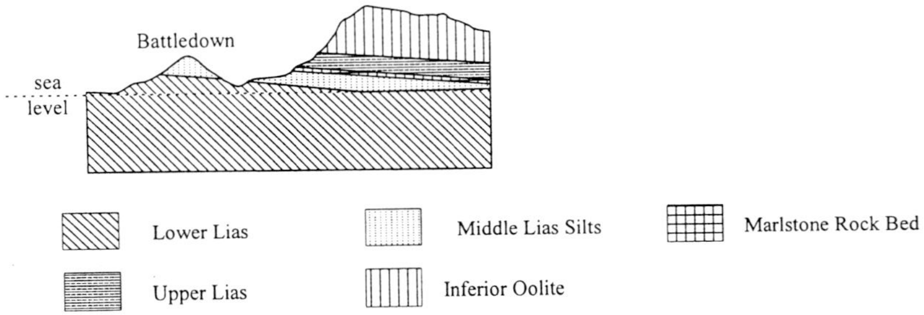

Section across Churchdown (after Murchison, 1844)

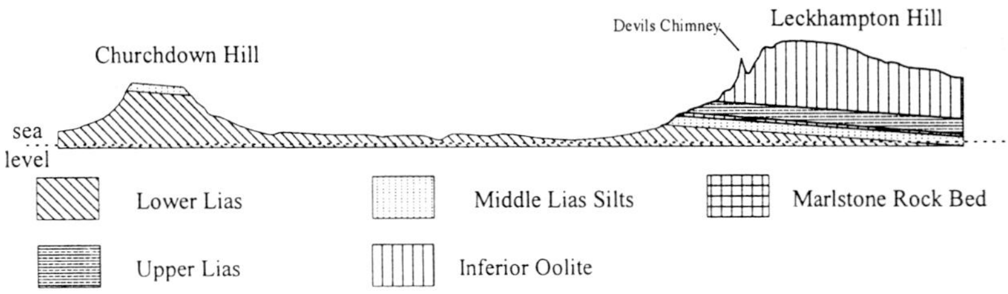

Section across Dundry Hill (after BGS)

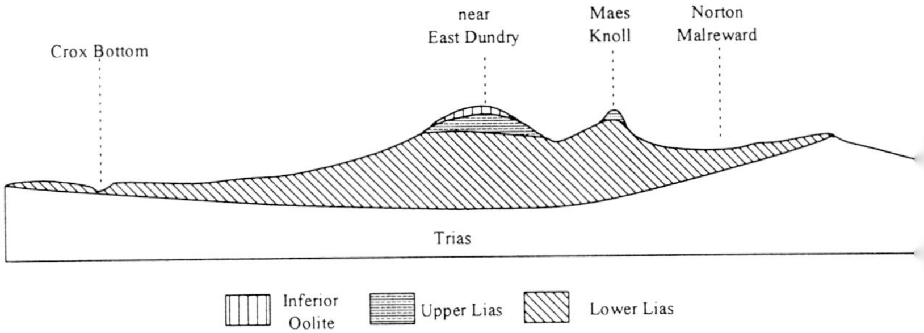

Chapter 7

7.1 Conclusions

The Cotswolds are a classic cuesta terrain developed on gently dipping Jurassic strata, which comprise both permeable limestones and less permeable clays and other sedimentary rocks. The juxtaposition of these different strata account for the form of the escarpment, the discharge characteristics of springs and dip slope rivers, and the location and nature of mass movement phenomena. While the area is by no means a classic karstic terrain, it has a fluviokarst (dry valley) network, some minor caves and solution hollows, and intermittent streams courses. Rates of limestone solutional loss are comparable to those in other parts of the UK. It is probable that two factors contribute to the paucity of true karstic forms-the relative incompetance of the limestones and the absence of overlying shales or sandstones. Tufa deposits, which result from the precipitation of calcium carbonate from springs, are of widespread occurrence in the area. The form of the Cotswold escarpment and its valleys was given classic status by W.M. Davis, but among British escarpments that of the Coswolds is notable for the relatively frequent occurrence of outliers. The area is also notable for having more recorded mass movement features, both active and relict, than any other part of England and Wales. Finally, while the extent, age and impact of glaciation in the area is still controversial, and the origin of the Plateau Drift, far from from settled, the area's river terraces, aeolian deposits, misfit streams and dry valleys give testament to the importance of Quaternary climatic changes in moulding the morphology of this distinctive region.

Acknowledgements

The authors would like to thank the following who have over the years studied the geomorphology of the Cotswolds: the late E.S. Hoare (Dean Close School, Cheltenham), Louis de Courcy Wheeler, Jenny Coe and Pat Butler (all of Hertford College, Oxford), and Mike Hart (of Keble College, Oxford). Thanks also go Mike Sumbler (British Geological Survey, Keyworth) for additional information on the locations of certain Quaternary deposits. Their researches have added greatly to our knowledge of the area. Finally thanks go to Lionel Walrond for comments on the text, to Peter Hayward for help with some of the diagrams and Martin Barfoot for the photographs.

Glossary

alluviation	The process by which material is deposited by running water.
anticline	A fold in geological strata that is convex upwards, i.e. forming an arch.
arenaceous	A term applied to sedimentary rocks composed of cemented sand.
consequent stream	A stream which flows in the direction of the original slope of the land surface.
cryoturbation	The process whereby soils, rock and sediments are churned up by frost processes, to produce convolutions or involutions.
cuesta	A ridge which possesses both scarp and dip slopes.
denudation	The laying bare of underlying rocks or strata by the removal of overlying material.
dip slope	The more gentle slope of a cuesta.
eutrophic	Water bodies which are bound in nutrients and therefore highly productive.
geomorphology	The subject which deals with the form of the ground surface and the processes which mould them.
fault	A crack or fissure in rock, the product of fracturing as a result of tectonic movement.
Flandrian	A term used to describe the Holocene in Britain, rather than globally.
Holocene	The second series of the Quaternary period, following the Pleistocene, often called the Post Glacial or Flandrian.
involution	A term synonymous with cryoturbation.
karst	Term given to limestone areas which contain topographically distinct scenery.
orogeny	The event or mechanism, of the construction of characteristically linear or arcuate mountain chains formed on continents.
periglacial	Climatic or geomorphic conditions of areas peripheral to ice sheets and glaciers.
Pleistocene	The first epoch of the Quaternary, preceeded by the Pliocene and succeeded by the Holocene.
Quaternary	The second period of the Cainozoic era. It followed the Tertiary and comprises the Pleistocene and Holocene.
scarp	The steep slope of a cuesta.
subsequent stream	A stream which follows a course determined by the structure of the local bed rock.
syncline	A trough in folded strata.
Tertiary	The first period of the Cainozoic era, comprising the Palaeocene through to the Pliocene.
tundra	Vast level, tree-less and mashy regions, usually with permanently frozen sub-soil.

References

Aalto, M.M., Coope, G.R and Gibbard, P.L (1984) Late-Devensian river deposits beneath the Floodplain Terrace of the River Thames at Abingdon, Berkshire, England. *Proceedings of the Geologists' Association*, **95**: 65-79

Ackerman, K.J and Cave, R (1967) Superficial deposits and structures including landslip, in the Stroud district, Gloucestershire. *Proceedings of the Geologists' Association*, **78**: 567-586

Adlam, B.H (1974) *The Worcester District 1: 50 000 Sheet 150: Worcester and the Malverns.* British Landscapes through Maps 17, Geographical Association, Sheffield

Ager, D.V., Donovan, D.T., Kennedy, W.S., McKerrow, W.S, Mudge, D.C and Sellwood, B.W (1973) The Cotswold Hills. *Geologists' Association Guide* No **36**, 34pp

Allen, T.G and Robinson, M.A (1993) *The Prehistoric and Iron Age Enclosed Settlement at Mingies Ditch, Hardwick-with-Yelford.* Thames Valley Landscapes: the Windrush Valley Vol **2**. Oxford Archæological Unit/Oxford University Commitee for Archæology

Anon (1925) Field meeting: Kenilworth Abbey. *Proceedings of the Cotteswold Naturalists' Field Club*, **XXII**: 77-78

Arkell, W.J (1931) The upper Great Oolite, Bradford Beds and Forest Marble of south Oxfordshire and the succession of gastropod faunas in the Great Oolite. *Quarterly Journal of the Geological Society of London*, **87**: 583-629

Arkell, W.J (1933) *The Jurassic System of Great Britain.* Clarendon Press, Oxford

Arkell, W.J (1945) Three Oxfordshire palaeoliths and their significance for Pleistocene correlation. *Proceedings of the Prehistoric Society*, **2**: 20-31

Arkell, W.J (1947a) *The Geology of Oxford.* Clarendon Press, Oxford, 268pp.

Arkell, W.J (1947b) The geology of the Evenlode Gorge, Oxfordshire. *Proceedings of the Geologists' Association*, **84**: 87-113

Arkell, W.J and Donovan, D. T (1951) The Fuller's Earth of the Cotswolds and its relation to the Great Oolite. *Quarterly Journal of the Geological Society,* **CVII** (3): 227-253

Aveline, W.T and Trench, R (1960) Geology of part of Northamptonshire. *Memoir of the the Geological Survey of Great Britain* (Old Series Sheet 53 SE), 19pp

Baker, P.G (1981) Interpretation of the Oolite Marl (Upper Aalenian, Lower Inferior Oolite) of the Cotswolds, England. *Proceedings of the Geologists' Association,* **92**: 169-188

Ballantyne, C.K and Harris, C (1994) *The Periglaciation of Great Britain.* Cambridge University Press, 330pp

Beckinsale, R.P (1934) *The Cotteswold Hills: A Geographical Interpretation.* Unpublished M.A. thesis. Univ. of London.

Beckinsale, R.P (1954) Geomorphology. In, Martin, A.F and Steel, R.W (ed's) *The Oxford Region: a Scientific and Historical Survey.* Oxford University Press

Beckinsale, R.P (1970) Physical problems of Cotswold rivers and valleys. *Proceedings of the Cotteswold Naturalists' Field Club*, **XXXV:** 194-205

Beckinsale, R.P (1982a) The Shill Brook: a study of an intermittent Cotswold river. *Proceedings of the Cotteswold Naturalists' Field Club*, **XXXVIII**: 71-80

Beckinsale, R.P (1982b) Pot-holes on the escarpment summit of the north Cotswolds. *Proceedings of the Cotteswold Naturalists' Field Club*, **XXXVIII**: 62-70

Beckinsale, R.P (1992) The Cotswold Hills: A Geographical Definition. *Proceedings of the Cotteswold Naturalists' Field Club*, **XL**(1): 20-30

Beckinsale, R.P and Beckinsale, M (1980) *The English Heartland.* Duckworth

Beckinsale, R.P and Beckinsale, R.D (1976) Glaciation and periglaciation in the Cotswolds. *Proceedings of the Cotteswold Naturalists' Field Club*, **XXXVII**: 46-57

Beckinsale, R.P and Richardson, L (1964) Recent findings of the physical development of the lower Severn valley. *Geographical Journal*, **130**: 87-105

Beckinsale, R.P and Smith, K.W (1955) Some morphological features of the valleys of the north Cotswolds. *Proceedings of the Cotteswold Naturalists' Field Club*, **XXXI:** 184-195

Beckit, H.O (1926) Physiography of the Oxford region. In, Walker, J.J (ed) *The Natural History of the Oxford District.* Oxord Universiry Press: 1-20

Bell, A.M (1894) On the Pleistocene gravels at Wolvercote near Oxford. *Report of the British Association, Oxford*: 663-664

Bell, A.M (1904) Implementiferous sections at Wolvercote (Oxfordshire). *Quarterly Journal of the Geological Society of London*, **60**: 120-132

Bennison, G.M and Wright, A.E (1969) *The Geological History of the British Isles.* Edward Arnold

Bishop, W.W (1958) The Pleistocene geology and geomorphology of three gaps in the Midland Jurassic escarpment. *Philosophical Transactions of the Roycl Society of London*, **B 241**:255-306

Boneham, B.F.W and Wyatt, R.J (1993) The stratigraphical position of the Middle Jurassic (Bathonian) Stonesfield Slate of Stonesfield, Oxfordshire, U.K. *Proceedings of the Geologists' Association*, **104**: 123-136

Bowen, D.Q., Rose, J., McCabe, A.M and Sutherland, D.G (1986) Correlation of Quaternary Glaciations in England, Ireland, Scotland and Wales. *Quaternary Science Reviews*, **5**: 299-340

Bravender, J (1867) On the watershed of the Upper Thames. *Geological Magazine*, **IV**: 422-424

Bravender, J (1868) The watershed of the Upper Thames. *Proceedings of the Cotteswold Naturalists' Field Club*, **IV**: 240-248

Bridgland, D. R (1988) Problems in the application of lithostratigraphic classification to Pleistocene river deposits. *Quaternary Newsletter*, **55**:1-8

Bridgland, D. R (1994) *Quaternary of the Thames*. Geological Conservation Review Series No 7, Chapman and Hall, London, 441pp.

Briggs, D. J (1973) *Quaternary deposits of the Evenlode valley and adjacent areas*. Unpublished Ph.D. thesis, University of Bristol

Briggs, D. J (1975) Origin, depositional environment and age of the Cheltenham Sand and Graval and related deposits. *Proceedings of the Geologists' Association*, **86**: 333-348

Briggs, D.J and Courteney, F.M (1972) Ridge and trough topography in the North Cotswolds. *Proceedings of the Cotteswold Naturalists' Field Club*, **XXXVI**: 94-103

Briggs, D.J and Gilbertson, D.D (1973) The age of the Hanborough Terrace of the River Evenlode, Oxfordshire. *Proceedings of the Geologists' Association*, **84**: 155-174

Briggs, D.J and Gilbertson, D.D (1974) Recent studies of Pleistocene deposits in the Evenlode valley and adjacent areas of the Cotswolds. *Sound*, **3**: 7-22

Briggs, D.J and Gilbertson, D.D (1980) Quaternary processes and environments in the Upper Thames basin. *Transactions of the Institute of British Geographers*, **5**: 53-65

Briggs, D.J *et al.* (1975) New interglacial site at Sugworth. *Nature*, **257:** 477-479

Briggs, D.J., Coope, G.R and Gilbertson, D.D (1975) Late Pleistocene terrace deposits at Beckford, Worcestershire, England. *Geological Journal*, **10**: 1-16

Briggs, D.J., Coope, G.R and Gilbertson, D.D (1985) *The Chronology and Environmental Framework of Early Man in the Upper Thames Valley*. British Archæological Series, **137**: BAR, Oxford

Briggs, D.J., Goudie, A.S and Hart, M.G (1976) Excursion to Upper Slaughter and the Cotswold scarp. In, Roe, D (ed) *A Field Guide to the Oxford Region*. Quaternary Research Association, Oxford: 53-60

Brodie, P.B (1850) Sketch map of the geology of the neighbourhood of Grantham, Lincolnshire. *Annual Magazine of Natural History*, **6**: 256-266

Bromehead, C. N (1909) Solution theory of valley formation. *Geological Magazine*, **6**: 141-142

Brown, E.H (1969) Jointing, aspect and the orientation of scarp-face dry valleys near Ivinghoe, Buckinghamshire. *Transactions of the Institute of British Geographers*, **46**: 61-73

Brunsden, D (1968) *Dartmoor*. British landscapes through maps, No **13,** Geographical Association, Sheffield

Buckland, W (1823) *Reliqiae Diluvianae: or Observation on the Organic Remains Contained in Caves, Fissures, and Diluvial Gravel and on Other Geological Phenomena, Attesting the Action of a Universal Deluge*. John Murray, London, 303pp

Buckman, S.S (1889) On the Cotteswold, Midford and Yeovil Sands, and the division between the Lias and the Chalk. *Quarterly Journal of the Geological Society of London*, **XLIX**: 445-458

Buckman, S.S (1895) The Bajocian of the Mid Cotteswolds. *Quarterly Journal of the Geological Society,* **LI**: 389

Buckman, S.S (1897) Deposits of the Bajocian age in the northern Cotteswolds: the Cleeve Hill plateau. *Quarterly Journal of the Geological Society of London*, **LV**: 220-223

Buckman, S.S (1899) The development of rivers: and particularly the genesis of the Severn. *Nat. Science*, **XIV**: 273-289

Buckman, S.S (1900) Excursion notes: chiefly on river features. *Proceedings of the Cotteswold Naturalists' Field Club*, **XIII**: 175-192

Buckman, S.S (1901) Bajocian and contiguous deposits in the North Cotteswolds. *Quarterly Journal of the Geological Society of London*, **LVII**: 140-149

Buckman, S.S (1903) The Cotteswold Hills: A Geographical Enquiry. *Proceedings of the Cotteswold Naturalists' Field Club*, **XIV**: 209-224

Bull, P and Goudie, A.S (1987) An examination of the ability of environmental reconstruction by SEM studies: a case study from the Plateau Drift deposits of Oxfordshire, England. In, Marshall, J.R (Ed) *Clastic Particles*. Van Nostrand Reinhold Company, New York: 36-50

Butler, P.B (1983) *Landsliding and Other Large Scale Mass Movements on the Escarpment of the Cotswold Hills*. Unpublished B.A. Dissertation, University of Oxford

Castleden, R (1977) Periglacial sediments in central and southern England. *Catena*, **4**: 111-121

Cave, R (1977) The Geology of the Malmesbury District. *Memoir of the Geological Society of Great Britain* Sheet 251. HMSO

Chandler, R.J., Kellaway, G.A., Skempton, A.W and Wyatt, R.J (1976) Valley slope sections in Jurassic strata near Bath, Somerset. *Philosophical Transactions of the Royal Society of London*, A **283**: 527-556

Chidlaw, N and Campbell, M.J.A (1988) Sedimentation in the Cornbrash Limestone Formation of the Cotswold Hills. *Proceedings of the Geologist's Association,* **99**: 27-42

Chorley, R.J (1965) The application of quantitative methods to geomorphology. In, Chorley, R.J and Haggett, P (eds) *Frontiers in Geographical Teaching*. Methuen, London: 148-163

Chorley, R.J and Morgan, M.A (1962) Comparison of morphometric features, Uncka Mountains, Tennessee and North Carolina and Dartmoor, England. *Bulletin of the Geological Society of America,* **73**: 17-34

Clifford, E.M (1936) Mollusca from the Barnwood gravels. *Proceedings of the Cotteswold Naturalists' Field Club*, **XXVI** (I): 102

Coe, J (1993) *A Palaeoenvironmental Reconstruction of Part of the Plateau Drift of Oxfordshire, Based on a Sedimentary Core from Nettlebed.* Unpublished B.A dissertation, University of Oxford

Conybeare, W.D (1829) On the valley of the Thames. *Proceedings of the Geological Society,* **I**: 145-149

Conybeare, W.D and Phillips, W (1822) *Outline of the Geology of England and Wales.* London

Coope, G.R (1976) Assemblages of fossil Coleoptera from the terraces of the Upper Thames, near Oxford. In, Roe, D (ed) *Field Guide to the Oxford Region.* Quaternary Research Association, Oxford: 20-22

Cope, J.C.W., Duff, K.L., Parsons, C.F., Torrens, H.S., Wimbledon, W.A and Wright, J.K (1980) A correlation of Jurassic rocks in the British Isles. Part II Middle and Upper Jurassic. *Geological Society of London Special Report*, **15**

Cox, A.H and Trueman, A.E (1920) Intra-Jurassic movements and the underground structure of the southern Midlands. *Geological Magazine*, **LVII**: 198-208

Corbel, J (1973) Erosion en terrain calcaire: vitesse d'érosion et morphologie . *Annales de Géographie,* **336**: 97-120

Davies, A.M (1923) The abandonment and entrenchment meanders: Wye, Evenlode, Cherwell, Thames. *Proceedings of the Geologists' Association*, **XXXIV**: 141-142

Davis, W.M (1895) The development of certain English rivers. *Geographical Journal*, **V**: 127-146

Davis, W.M (1899) The drainage of cuestas. *Proceedings of the Geologists' Association*, **XVI**: 87-93

Davis, W.M (1909) The valleys of the Cotswold Hills. *Proceedings of the Geologists' Association*, **XXI**: 150-152

De Courcy Wheeler, L (1974) *Aspects of the Morphometry of the Oxford Region.* Unpublished B.A. dissertation, University of Oxford

De la Beche, H.T (1839) Report on the geology of Cornwall, Devon and west Somerset. *Memoirs of the Geological Survey*, London

Dimbleby, G.W and Evans, J.G (1974) Pollen and land-snail analysis of calcareous soils. *Journal of Archaeological Science*, **1**: 117-133

Dines, H.G (1946) Pleistocene and Recent deposits In, Richardson, L., Arkell, W.J and Dines, H.G (Ed's) *The Geology of the Country around Witney.* Mem. Geol. Surv. H.M.S.O

Douglas, I (1964) Intensity and periodicity in denudation processes with special reference to the removal of material in solution by rivers. *Zeitschrift für Geomorphologie*, NF **8**: 453-473

Douglas, J.A and Arkell, W.J (1928) The stratigraphical distribution of the Cornbrash I. the south-west area. *Quarterly Journal of the Geological Society of London*, **84**: 117-178

Dreghorn, W (1967) *Geology Explained in the Severn Vale and Cotswolds.* David and Charles: Newton Abbot, 187pp

Duigan, S.L (1956) Interglacial plant remains from the Wolvercote Channel, Oxford. *Quarterly Journal of the Geological Society of London,* **112** 363-372

Duncan, C.C and Richardson, L (1924) Hen Wood petrifying spring, Edgeworth. *Proceedings of the Cotteswold Naturalists' Field Club*, **XXII** (I): 55-56

Dury, G.H (1951) A 400-foot bench in south-eastern Warwickshire. *Proceedings of the Geologists' Association*, **62**: 167-173

Dury, G.H (1955) Meanders of the Coln valley: a new interpretation. *Proceedings of the Cotteswold Naturalists' Field Club*, **XXXI**: 206-219

Dury, G.H (1958) Tests of a general theory of misfit streams. *Transactions of the Institute of British Geographers*, **25**: 105-118

Dury, G.H (1965) Theoretical implications of underfit streams. *U.S.G.S. Professional Paper*, **452-A**

Edmonds, E.A., Poole, E.G and Wilson, V (1965) The Geology of the Country around Banbury and Edge Hill. *Memoir of the British Geological Survey*, Sheet 201 (England and Wales). H.M.S.O

Emery, F (1974) *The Oxfordshire Landscape.* Hodder and Stoughton

Evans, J.G (1972) *Land Snails in Archaeology.* Seminar Press, London

Evans, J.G and Jones, H (1973) Subfossil and modern land-snail faunas from rock-rubble habitats. *Journal of Conchology*, **28**: 103-129

Findlay, D.C (1976) Soils of the southern Cotswolds and surrounding country. *Memoirs of the Soil Survey of Great Britain, Sheets 251 and 265*

Forster, A., Hobbs, P.R.N., Monkhouse, R.A and Wyatt, R.J (1985) Environmental geology study; parts of west Wiltshire and south-east Avon. *British Geological Survey, Environmental Geology Study*

Gardiner, C.I (1932) Recent discoveries in the Stroud valley. *Proceedings of the Cotteswold Naturalists' Field Club*, **XXIV** (3): 163-170

Gardiner, C.I (1933) The occurrence of elk and other mammalia in the alluvium of the Stroud and Nailsworth valleys. *Proceedings of the Cotteswold Naturalists' Field Club*, **XXV** (1): 77-82

Gardiner, C.I (1938) The making of the valley of the Stroud Frome and an account of its fluviatile deposits. *Proceedings of the Cotteswold Naturalists' Field Club*, **XXVI** (3): 265-287

Gerrard, J and Morris, L (1980) Mass movement forms and processes on Bredon Hill, Worcestershire. *Department of Geography, University of Birmingham Working Paper Series*, **10**: 1-10

Gibbard, P.L (1985) *The Pleistocene History of the Middle Thames Valley.* Cambridge University Press, 155pp

Goudie, A.S (1967) *Solutional rates, processes and forms in the country between Cheltenham and Stow, Glos.* Unpublished B.A. dissertation, University of Cambridge

Goudie, A.S (1976) Introduction: the Oxford region. In, Roe, D (ed) *Field Guide to the Oxford Region.* Q.R.A, Oxford: 1-5

Goudie, A.S (1983) *Environmental Change* (2nd ed). Clarendon Press, Oxford

Goudie, A.S and Hart, M.G (1975) Pleistocene events and forms in the Oxford region. In, Smith, C.G and Scargill, D.I (eds) *Oxford and Its Region.* Oxford University Press: 3-13

Goudie, A.S and Hart, M.G (1976) The Cleeve Hill slip troughs. In, Roe, D (ed) *Field Guide to the Oxford Region.* Q.R.A, Oxford: 55-60

Goudie, A.S., Day, M.J and Hart, M.G (1980) Developments in the Geomorphology of the Oxford region. In, Rowley, T (ed*) The Oxford Region.* Oxford University Department of External Studies: 11-22

Goudie, A.S., Viles, H.A and Pentecost, A (1993) The late-Holocene tufa decline in Europe. *The Holocene*, **3** (2): 181-186

Green, G.W (1992) *British Regional Geology: Bristol and Gloucester Region 3rd ed.* HMSO, London, 188pp

Green, G.W and Melville, R.V (1956) The stratigraphy of the Stowell Park Borehole. *Bulletin of the Geological Survey of Great Britain*, **11**: 1-66

Gregory, K.J (1966) Aspects of landforms in N-E Yorkshire. *Biuletyn Peryglacjalny,* **15**: 115-119

Groom, G.E and Williams, V.H (1965) The solution of limestone in south Wales. *Geographical Journal*, **131**: 37-41

Grove, A.T (1953) Account of a mudflow on Bredon Hill, Worcestershire, April 1951. *Proceedings of the Geologists' Association*, **64**: 10-13

Harris, C (1987) Mechanisms of mass movement in periglacial environments. In, Anderson, M.G and Richards, K.S (eds) *Slope Stability*. Wiley, Chichester: 531-559

Harries, W.J.R (1977) *The sand and gravel resources of the Thames Valley, the country around Eynsham.* Mineral Assessment Report 28, Institute of Geological Sciences. London. H.M.S.O

Hart, M.G (1976) *Selected Aspects of the Fluvial Geomorphology of Some Cotswold Rivers.* Unpublished D. Phil. thesis, University of Oxford

Hawkins, A.B and Kellaway, G.A (1971) Field meeting at Bristol and Bath with special reference to new evidence of glaciation. *Proceedings of the Geologists' Association*, **82**: 267-292

Hawkins, A.B and Privett, K.D (1979) Engineering geomorphological mapping as a technique to elucidate areas of superficial structures; with examples from the Bath area of the south Cotswolds. *Quarterly Journal of Engineering Geology*, **12**: 221-233

Hawkins, A.B and Privett, K.D (1981) A building site on cambered ground at Radstock, Avon. *Quarterly Journal of Engineering Geology,* **14**: 151-167

Hawkins, A.B., Lawrence, M.S and Privett, K.D (1988) Implications of weathering on the engineering properties of the Fuller's Earth formation. *Géotechnique,* **34** (4): 517-532

Hazleton, J and Jarvis, M.G (1979) Age and significance of alluvium in the Windrush, Oxfordshire. *Nature*, **282**: 291-292

Hey, R.W (1965) Highly quartzose pebble gravels in the London Basin. *Proceedings of the Geologists' Association*, **76**: 403-420

Hey, R (1986) A re-examination of the Northern drift of Oxfordshire. *Proceedings of the Geologists' Association,* **97**: 291-301

Hollingworth, S.E., Taylor, J.H and Kellaway, G.A (1944) Large scale superficial structures in the Northampton Ironstone field. *Quarterly Journal of the Geological Society,* **99-100**: 1-44

Horton, A (1977) The age of the Middle Jurassic 'White Sands' of Oxfordshire. *Proceedings of the Geologists' Association*, **88**: 147-162

Horton, A (1988) Geological notes and local details for 1: 10 000 sheet SP20SW (Filkins). *British Geological Survey Technical Report* WA/88/31

Horton, A., Poole, E.G., Williams, B.J., Illing, V.C and Hobson, G.D (1988) Geology of the country around Chipping Norton. *Memoir of the British Geological Survey*, Sheet 218 (England and Wales)

Horton, A., Shepherd-Thorn, E.R and Thurrel, I.R (1974) The geology of the new town of Milton Keynes (explanation of sheet 1:25 000 special geological sheet SP 83 with parts SP 73, 74, 84, 93 and 97). *Report of the Institute of Geological Sciences*, **74/10**, 102pp

Huddleston, W.H (1878) Excursion to Chipping Norton. *Proceedings of the Geologists' Association*, **5**: 378-389

Hull, E (1855) On the Physical Geography and Pleistocene Phænomena of the Cotteswold Hills. *Quarterly Journal of the Geological Society*, **XI**: 482

Hull, E (1857) The Geology of the Country around Cheltenham. *Memoir of the Geological Survey*, Sheet 44

Hunt, C.O (1984) Palynology of the Cheltenham Gravels. In, Roberts, C.R (ed) *Geomorphology of the Lower Severn Area: Field Excursion Guide*. British Geomorphological Research Group, Cheltenham: 11-12

Hutchinson, J.N (1967) The free degradation of London Clay cliffs. *Proceedings of the Geological Conference Oslo*, **1**: 113-118

Hutchinson, J.N (1991) Periglacial and slope processes. In, Forster, A., Culshaw, M.G., Cripps, J.C., Little, J.A and Moon, C.F (ed's) *Quaternary Engineering Geology*. Geological Society Engineering Geology Special Publication No. **7**: 283-331

Jones, D.C.K and Lee, E.M (1994) *Landsliding in Great Britain*. HMSO, 361pp

Judd, J.W (1875) The geology of Rutland and parts of Lincoln, Leicester, Huntingdon and Cambridge. *Memoir of the Geological Survey of Great Britain*, 320pp

Kellaway, G, A (1991) *Hot Springs of Bath: Investigations of the Thermal Waters of the Avon Valley*. Bath City Council, Bath

Kellaway, G.A., Horton, A and Poole, E.G (1971) The development of some Pleistocene structures in the Cotswolds and Upper Thames basin. *Bulletin of the Geological Survey of Great Britain*, **37**: 1-28

Kellaway, G, A and Taylor, J.H (1968) The influence of landslipping on the development of the city of Bath, England. *Report of the International Geological Congress 23rd session, Czechoslovakia*, **12**: 65-76

Kerney, M.P., Preece, R,C and Turner, C (1980) Molluscan and plant biostratigraphy of some Late Devensian and Flandrian deposits in Kent. *Philosophical Transactions of the Royal Society of London*, **B 291**: 1-43

Knight, H.H and Riddelsdell, H.J (1933) Excursion to Whittington, Sevenhampton Common and Watercombe Marsh, Brimpsfield. *Proceedings of the Cotteswold Naturalists' Field Club,* **XXV** (I): 14-15

Lake, P (1915) *Physical Geography.* Cambridge University Press

Lambrick, G and Robinson, M.A (1988) The development of the floodplain grassland in the Upper Thames Valley. In, Jones, M (Ed) *Archæology and the Flora of the British Isles.* Oxford University Commitee for Archæology Monograph **14**: 55-75

Lapworth, H (1911) The geology of dam trenches. *Transactions of the Institution of Water Engineers*, **16**: 25-66

Large, N.F (1963) The superficial deposits of the Stroud valley. *Proceedings of the Cotteswold Naturalists' Field Club*, **XXXIII**: 192-197

Large, N.F and Sparks, B.W (1961) The non-marine mollusca of the Cainscross Terrace, near, Stroud, Gloucestershire. *Geological Magazine*, **98**: 423-426

Lucy, W.C (1872) The gravels of the Severn, Avon and Evenlode, and their extension over the Cotteswolds Hills. *Proceedings of the Cotteswold Naturalists' Field Club*, **V**: 71-142

Lucy, W.C (1880) On the extension of the Northern Drift and Boulder Clay over the Cotteswold range. *Proceedings of the Cotteswold Naturalists' Field Club*, **VII**: 50-61

Mackintosh, D (1869) *The Scenery of England and Wales.* Longmans, Green and Co, London

Macrae, R.J (1985) Palaeolithic archaeology of the Upper Thames basin. In, Briggs, D.J., Coope, G.R and Gilbertson, D.D (eds) *The Chronology and Environmental Framework of Early Man in the Upper Thames Valley.* British Archæological Series, **137**: BAR, Oxford

Maddy, D., Lewis, S.G and Green, C.P (1991) A review of the stratigraphic significance of the Wolvercote Terrace of the Upper Thames. *Proceedings of the Geologists' Association*, **102** (3): 217-225

Martin, A.F and Steel, R.W (eds) (1954) *The Oxford Region: a Scientific and Historical Survey.* Oxford University Press

McKerrow, W.S and Baker, S (1988) Field meeting to Charlbury and Stonesfield, Oxfordshire. *Proceedings of the Geologist's Association,* **99**: 61-65

McKerrow, W.S and Kennedy, W.J (1973) *Geologist's Association Guides No3: The Oxford District.* Benham and Co: Colchester

Metcalf, S.J., Vaughan, R. F., Benton, M.J., Cole,J., Simms, M.J and Dartnall, D.L (1992) A new Bathonian (Middle Jurassic) microinvertebrate site, within the Chipping Norton Limestone Formation at Hornsleasow Quarry, Gloucestershire. *Proceedings of the Geologist's Association,* **103**: 321-342

Miller, H (1887) The country around Otterburn and Elsden. *Memoir of the Geological Survey of Great Britain* (England and Wales)

Mitchell, G.F., Penney, L.F., Shotton, F.W and West, R.G (1973) A correlation of Quaternary deposits in the British isles. *Geological Society of London Special Report* No **4,** 99pp

Morgan, R.P.C (1971) A morphometric study of some valleys in the English Chalklands. *Transactions of the Institute of British Geographers*, **54**: 33-44

Morris, L (1974) The geomorphology of Bredon Hill. In, *Worcester and its Region: Field Studies in the Former County of Worcestershire*. Geographical Association, Worcester Branch: 174-183

Mudge, D.C (1978) Stratigraphy and sedimentation of the Lower Inferior Oolite of the Cotswolds. *Journal of the Geological Society of London*, **135**: 611-627

Murchison, R. (1834) Outline of the *Geology of the Neighbourhood of Cheltenham.* H. Davies, Cheltenham

Murchison, R. I (1844) Outline of the *Geology of Cheltenham* new ed, augmented and revised by Strickland, H.E and Buckman, J H. Davies, Cheltenham; John Murray London

Murray, J.W and Hawkins, A.B (1973) Geology and physical environment. In, Hadfield, C and Hadfield, M (ed's) *The Cotswolds - a New Study*. David and Charles, Newton Abbot: 21-47

O'Neill, H.E and Shotton, F.W (1974) Mammoth remains from gravel pits in the north Cotswolds. *Proceedings of the Cotteswold Naturalists' Field Club*, **36** (4): 196-197

Parker, A.G (1995) *Late-Quaternary Environmental Change in the Upper Thames Basin, Central-Southern England.* Unpublished D.Phil. Thesis, University of Oxford.

Parks, C.D (1991) A review of the possible mechanisms of cambering and valley bulging. In, Forster, A., Culshaw, M.G., Cripps, J.C., Little, J.A and Moon, C.F (ed's) *Quaternary Engineering Geology*. Geological Society Engineering Geology Special Publication No. 7: 373-380

Paterson, K (1970) *Aspects of the Geomorphology of the Oxford Region*. Unpublished D.Phil. Thesis, University of Oxford,

Perrin, R.M.S (1963) *Studies in Pedogenesis.* Unpublished Ph.D thesis, University of Cambridge

Phillips, J (1871) *Geology of Oxford and the Valley of the Thames*. Clarendon Press, Oxford

Pitty, A.F (19680 The scale and significance of solutional loss from the limestone tract of the southern Pennines. *Proceedings of the Geologists' Association*, **LXXIX**: 153-177

Playne, G.F (1868) Notes on the landslips near Nailsworth. *Proceedings of the Cotteswold Naturalists' Field Club*, **IV**: 230-232

Playne, G.F (1869) On the physical geography of the district drained by the River Frome and its tributaries. *Proceedings of the Cotteswold Naturalists' Field Club*, **V**: 21-38

Playne, G.F (1874) On the Recent calcareous deposits of the Cotteswolds. *Proceedings of the Cotteswold Naturalists' Field Club*, **VI**: 81-89

Ramsay, A.C., Aveline, W.T and Hull, E (1885) The geology of parts of Wiltshire and Gloucestershire (Sheet 34). *Memoir of the Geological Survey of Great Britain*

Rastall, R.H (1925) On the tectonics of the southern Midlands. *Geological Magazine*, **LXXII**: 193-222

Rastall, R.H (1926) Note on the geology of the Bath springs. *Geological Magazine*, **63**: 98-104

Richardson, L (1902) Notes of the geology of Bredon Hill. *Transactions of the Woolhope Naturalists' Field Club* for 1902 : 62-67

Richardson, L (1904a) *A Handbook to the Geology of Cheltenham*. Norman Sawyer & Co, Cheltenham

Richardson, L (1904b) Bredon Hill: its geologic structure and water resources. *Transactions of the Worcestershire Naturalists' Club*, **VIII**: 82-88

Richardson, L (1908) The geology of Ebrington Hill, north Cotteswolds. *Proceedings of the Cotteswold Naturalists' Field Club*, **XVI**: 129-134

Richardson, L (1911 a) The Inferior Oolite and contiguous deposits of the Chipping Norton district, Oxfordshire. *Proceedings of the Cotteswold Naturalists' Field Club*, **XVII** (2): 202

Richardson, L (1912) Memoir explanatory of a map of a part of Cheltenham. *Proceedings of the Cotteswold Naturalists' Field Club*, **XVII**: 297-319

Richardson, L (1913a) Memoir explanatory of a map of a part of Cheltenham. *Proceedings of the Cotteswold Naturalists' Field Club*, **XVIII** (2): 125-136

Richardson, L (1913b) A 'Swallow Hole' in the Inferior Oolite near Cheltenham. *Proceedings of the Cotteswold Naturalists' Field Club*, **XVIII**: 66

Richardson, L (1923a) Certain Jurassic (Aalenian-Vesulian) strata of southern Northamptonshire. *Proceedings of the Geologists' Association*, **34**: 97-113

Richardson, L (1923b) Eyford Park. *Proceedings of the Cotteswold Naturalists' Field Club*, **XXI** (2): 74-76

Richardson, L (1923c) Thames Head. *Proceedings of the Cotteswold Naturalists' Field Club*, **XXI** (2): 161-166

Richardson, L (1924) Excursion to Calmsden, Winterwell and Ampney Crucis. *Proceedings of the Cotteswold Naturalists' Field Club*, **XXI** (3): 183-185

Richardson, L (1925a) Certain Jurassic (Aalenian-Vesulian) strata of southern Northamptonshire. *Proceedings of the Cotteswold Naturalists' Field Club*, **XXII**: 137-152

Richardson, L (1925b) The River Leach. *Proceedings of the Cotteswold Naturalists' Field Club*, **XXII**(1): 43-50

Richardson, L (1926a) The River Dikler. *Proceedings of the Cotteswold Naturalists' Field Club,* **XXII** (2): 179-181

Richardson, L (1926b) The Sherbourne Brook. *Proceedings of the Cotteswold Naturalists' Field Club*, **XXII** (2): 183-185

Richardson, L (1927 a) Excursion to Wotton-under-Edge, Boxwell and Dursley. *Proceedings of the Cotteswold Naturalists' Field Club*, **XXII** (3): 205-206

Richardson, L (1927 b) The River Eye. *Proceedings of the Cotteswold Naturalists' Field Club*, **XXII** (3): 255-258

Richardson, L (1927c) The Dursley tufa. *Proceedings of the Cotteswold Naturalists' Field Club*, **XXII** (3): 267-270

Richardson, L (1929) The Country Around Moreton In Marsh. *Memoirs of the Geological Survey of England and Wales*, Sheet **217**, HMSO, pp162

Richardson, L (1930) *Wells and Springs of Gloucestershire.* Memoirs of the Geological Survey Research, HMSO

Richardson, L (1933) The Country Around Cirencester. *Memoirs of the Geological Survey of England and Wales*, Sheet **235**, HMSO, pp119

Richardson, L (1935) Week-end field meeting in the Witney district. *Proceedings of the Geologists' Association*, **XLVI**: 403-411

Richardson, L (1937a) Moreton Valley. *Proceedings of the Cotteswold Naturalists' Field Club*, **26**:

Richardson, L (1937b) Toadsmoor pond springs, Bisley. *Proceedings of the Cotteswold Naturalists' Field Club*, **26**:

Richardson, L (1941) A swallow hole near Cheltenham. *British Caver*, **VIII:**

Richardson, L (1955) The flint gravel on Bathampton Down, Bath. *Proceedings of the Cotteswold Naturalists' Field Club*, **32** (2): 41

Richardson, L (1972*) A Handbook to the Geology of Cheltenham and Neighbourhood.* Revised by R.D. Beckinsale. Paul P.B. Minet, Chicheley

Richardson, L and Sandford, K.S (1960) Great Chessels gravel pit near Bourton-on-the-Water, Glos., and the occurrence of mammalian remains. *Proceedings of the Geologists' Association*, **71**: 40-46

Richardson, L and Sandford, K.S (1963a) Ditchford gravel pit near Stretton-on-Fosse, Gloucestershire, and the occurrence of a mammoth tooth. *Proceedings of the Cotteswold Naturalists' Field Club*, **33** (4): 172-176

Richardson, L and Sandford, K.S (1963b) Find of a mammoth tooth at Charlton Kings, Cheltenham. *Proceedings of the Cotteswold Naturalists' Field Club*, **33** (4): 170-171

Richardson, L, Arkell, W.J and Dines, H.G (1946) The Country Around Witney. *Memoirs of the Geological Survey of England and Wales*, Sheet **236**, HMSO, pp162

Robinson, M.A (1981) *Investigations of Palaeoenvironments in the Upper Thames Valley, Oxfordshire.* Unpublished Ph.D thesis, University of London

Robinson, M.A and Lambick, G (1984) Holocene alluviation and hydrology in the Upper Thames basin. *Nature,* **308**: 809-814

Robinson, M.A and Wilson, B (1987) A survey of the environmental archaeology in the south Midlands. In Keeley, H.C.M (Ed) *Environmental Archaeology: A regional review, Vol II.* Historic Buildings and Monuments Commision for England. Occasional Paper No **1**: 16-84

Robson, P (1976) The sand and gravel resources of the Thames valley, the country between Lechlade and Standlake. *Mineral Assessment Report* **23**. Insitute of Geological Sciences, London. H.M.S.O

Roe, D (1981) *The Lower and Middle Palaeolithic Periods in Britain.* Routledge and Kegen Paul: London

Roe, D (Ed) (1976) *Field Guide to the Oxford Region.* Quaternary Research Association, Oxford

Rose, J (1987) Status of the Wolstonian glaciation in the British Quaternary. *Quaternary Newsletter*, **53**: 1-9

Sandford, K.S (1924) The gravels of the Oxford district. *Quarterly Journal of the Geological Society of London*, **80**: 113-176

Sandford, K.S (1926) Pleistocene deposits. In, Pringle, J (ed) The Geology of the Country Around Oxford. *Memoir of the Geological Survey of Great Britain*, (England and Wales) H.M.S.O: 104-172

Sandford, K.S (1929) The erratic blocks and the age of the southern limit of glaciation in the Oxford district. *Quarterly Journal of the Geological Society of London*, **85**: 359-388

Sandford, K.S (1932) Some contributions to the Pleistocene succession in England. *Geological Magazine*, **69**: 1-17

Sandford, K.S (1954) River development and superficial deposits. In, Martin, A.F and Steel, R.W (eds) (1954) *The Oxford Region: a Scientific and Historical Survey.* Oxford University Press

Sandford, K.S (1965) Notes on the gravels of the Thames between Lechlade and Dorchester. *Proceedings of the Geologists' Association*, **76**: 61-76

Sandford, K.S and Richardson, L (1961) Mammoths in the Cotswolds. *Nature*, **190**: 342-343

Seddon, M.B and Holyoak, D.T (1985) Evidence of sustained regional permafrost during deposition of fossiliferous Late Pleistocene river sediments at Stanton Harcourt (Oxfordshire, England). *Proceedings of the Geologists' Association,* **96**: 53-71

Sellwood, D.W and McKerrow, W.S (1974) Depositional environments in the lower part of the Great Oolite of Oxfordshire and north Gloucestershire. *Proceedings of the Geologists' Association,* **85**: 199-220

Shackleton, N.J and Opdyke, D.D (1973a) Quaternary isotope and palaeomagnetic stratigraphy of equiatorial Pacific core V28-238: oxygen isotope temperatures and ice volume on a 10^5 and 10^6 year scale. *Quaternary Research*, **3**: 39-55

Shackleton, N.J and Opdyke, D.D (1973b) Oxygen isotope and palaeomagnetic stratigraphy of Pacific core V28-239, Late Pliocene to latest Pleistocene. In, Cline, R.W and Hays, J.D (eds) Investigation of Late Quaternary palaeogeography and palaeoecology. *Geological Society of America Memoir* **145**, Boulder: 449-464

Shotton, F.W (1953) The Pleistocene deposits of the area between Coventry, Rugby and Leamington and their bearing on the topographical development of the Midlands. *Philosophical Transactions of the Royal Society of London,* **B237**: 209-260

Shotton, F.W (1960) Large-scale patterned ground in the valley of the Worcestershire Avon. *Geological Magazine*, **97**: 404-408

Shotton, F.W (1973) English Midlands. In, Mitchell, G.F., Penney, L.F., Shotton, F.W and West, R.G (ed's) A correlation of Quaternary deposits in the British isles. *Geological Society of London Special Report* No **4**:

Shotton, F.W., Blundel, P.J and Williams, R.E.G (1970) Birmingham University radiocarbon dates IV. *Radiocarbon*, **12**: 141-156

Shotton, F.W., Goudie, A.S., Briggs, D.J and Osmaston, H.A (1980) Cromerian interglacial deposits at Sugworth, near Oxford, England, and their relation to the Plateau Drift of the Cotswolds and terrace sequence of the upper and middle Thames. . *Philosophical Transactions of the Royal Society of London*, **289B:** 55-86

Small, R.J (1970) *The Study of Landforms.* Cambridge University Press, Cambridge.

Smith, D.I (1965) Some aspects of limestone solution in the Bristol region. *Geographical Journal*, **131**: 44-49

Smith, C.G and Scargill, D.I (eds) (1974) *Oxford and its Region: Geographical Essays*. Oxford University Press

Strahler, A.N (1952) Dynamic basis of geomorphology. *Bulletin of the Geological Socity of America,* **63**: 923-936

Sumbler, M.G (1983) A new look at the type Wolstonian glacial deposits of central England. *Proceedings of the Geologists' Association,* **94**: 23-31

Sumbler, M.G (1984) The stratigraphy of the Bathonian White Limestone and Forest Marble Formations of Oxfordshire. *Proceedings of the Geologists' Association,* **95**: 57-64

Sumbler, M.G (1994) Geological notes and local details for 1: 10 000 sheet SP21SW (The Barringtons). *British Geological Survey Technical Report* WA/94/92

Sumbler, M.G (1995a) The terraces of the rivers Thame and Thames, and their bearing on the chronology of glaciation in central and eastern England. *Proceedings of the Geologists' Association,* **106**: 93-106

Sumbler, M.G (1995b) Geological notes and local details for 1: 10 000 sheet SP00SE (The Ampneys). *British Geological Survey Technical Report* WA/95/74

Sumbler, M.G. and Barron, A.J.M.(1996 in press) The type section of the Hampen Formation (Middle Jurassic, Great Oolite Group) at Hampen Cutting, Gloucestershire. *Proceedings of the Cotteswold Naturalists' Field Club,* **XLI**: 000-000

Sumbler, M.G *et al.* (in prep) Geology of the Country around Cirencester *Memoir of the British Geological Survey* (England and Wales) Sheet 235

Sweeting, M.M (1972) *Karst Landforms*. Macmillan

Sweeting, M.M (1965) Denudation in limestone regions: an introduction *Geographical Journal*, **131**: 34-37

Thompson, B (1910) Northamtonshire (including contiguous parts of Rutland and Warwickshire) geology in the field. *Geological Association of London Jubilee Volume*: 450-487

Tomlinson, M.E (1929) The drifts of the Stour-Evenlode watershed and their extension into the valleys of the Warwickshire Stour and Upper Evenlode. *Proceedings of the Birmingham Natural History and Philosophical Society*, **15**: 157-169

Tomlinson, M.E (1941) Pleistocene gravels of the Cotswold sub-edge plain from Mickleton to the Frome Valley. *Quarterly Journal of the Geological Society of London*, **96**(4): 386-420

Tomlinson, M.E (1963) The Pleistocene Chronology of the English Midlands. *Proceedings of the Geologists' Association*, **74**: 187-202

Townsend, J (1813) *The Character of Moses Established for Veracity as an Historian, Recording Events from the Creation to the Deluge*. London

Tydelsley, J (1986,a) *The Wolvercote Channel Handaxe Assemblage: A comparative Study.* British Archæological Report, British Series, 153,211pp

Tydelsley, J (1986,b) A re-assessment of the handaxe assemblage recovered from the Wolvercote Channel, Oxford. In, Colcutt, S.N (Ed) *Recent Studies in the Palaeolithic of Britain and its Nearest Neighbours.* J.R. Collis Publications, Department of Archæology and Prehistory, Sheffield University: 23-25

Tydelsley, J (1988) Quartzite implements recovered from the Wolvercote Channel Oxfordshire. In, Macrae, R.J and Moloney, N (Ed's) *Non-flint stone tools and the Palaeolithic occupation of Britain.* British Archæological Report, British Series, 189:159-166

Varney, W.D (1921) The geological history of the Pewsey Vale. *Proceedings of the Geologists' Association,* **XXXII**: 198-205

Viles, H.A and Goudie, A.S (1990) Tufas, travertines and allied carbonate deposits. *Progress in Physical Geography,* **14**: 19-41

Walford, A.E (1878) On some Middle and Upper Lias beds in the neighbourhood of Banbury. *Proceedings of the Warwickshire Naturalists' and Archaeological Field Club* suppl. for 1878: 1-23

Walford, A.E (1906) *On Some New Oolitic Strata in North Oxfordshire.* Buckingham, Privately Printed, 32pp

Watts, W (1905) Geological notes on sinking Langsett and Underbank concrete trenches in the Little Don valley. *Transactions of the Institute of Mining Engineers,* **31**: 668-680

White, H.J., Osbourne (1897) On the origin of the high-level gravel with triassic debris adjoining the valley of the Upper Thames. *Proceedings of the Geologists' Association,* **XV**: 157-174

Whiteman, C.A (1991) Broadening the Thames - some thoughts on stratigraphic procedure. *Quaternary Newsletter,* **64**: 19-28

Whittaker, A (1974) Geology of Bredon Hill, Worcestershire. *Bulletin Geological Survey of Great Britain,* **42**: 1-49

Wilkinson, K.N., Berzins, V., Branch, N.P., Fairburn, A.S and Lowe, J.J (1994) *Great Rissington/Bourton Link Sewerage scheme: a study of the Late Devensian and early to middle Flandrian environments from material recorded during the watching brief.* Report for Thames Water Ltd/Cotswold Archaeological Trust, 38pp

Williams, B.J and Whittaker, A (1974) The geology of the country around Stratford-upon Avon and Evesham. *Memoir of the Geological Survey of Great Britain.* 125pp

Williams, P.W(1963) An initial estimate of the speed of limestone solution in County Clare. *Irish Geog.,* **4**: 432-441

Williamson, I.T (1995) Geological notes and local details for 1: 10 000 sheet SP21NW (Fifield). *British Geological Survey Technical Report* WA/95/79

Willing, M. J (1985) *The Biostratigraphy of Flandrian Tufa Deposits in the Cotswold and Mendip Hills*. Unpublished Ph.D thesis, University of Sussex

Wills, L.J (1938) The Pleistocene development of the Severn from Bridgnorth to the sea. *Quarterly Journal of the Geological Society of London*, **XCIV**: 161-242

Witchell, E (1867) On the denudation of the Cotswolds. *Proceedings of the Cotteswold Naturalists' Field Club*, **IV**: 214-230

Witchell, E (1877) On the angular gravel of the Cotteswolds. *Proceedings of the Cotteswold Naturalists' Field Club*, **VI**: 146

Witchell, E (1882) *The Geology of the Stroud and the Area Drained by the Frome*. H.James, Stroud

Witchell, E (1886) On the Basement Beds of the Inferior Oolite of Gloucestershire. *Quarterly Journal of the Geological Society of London*, **XLII**: 264-271

Worssam, B.C and Bisson, G (1961) The geology of the country between Sherbourne, Gloucestershire and Burford, Oxfordshire. *Bulletin of the Geological Survey of Great Britain*, **17**: 75-115

Worssam, B.C., Ellison, R.A and Moorlock, B.S.P (1989) Geology of the country around Tewkesbury. *Memoir of the British Geological Survey*, sheet 216 (England and Wales)

Wymer, J.J (1968) *Lower Palaeolithic archaeology in Britain as represented by the Thames Valley*. John Baker: London